skin

_)_)_)_
• • • •
 • • •

the body literacy library

skin

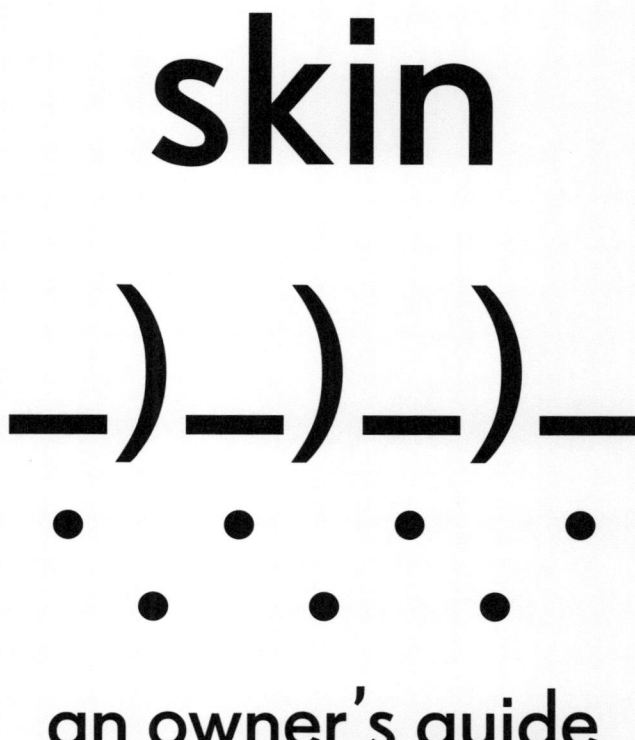

an owner's guide

Dr Emma Craythorne

the body literacy library

Body literacy is a human right. It is a means to observe,
learn, and understand ourselves – three essential steps to
enhancing our self-knowledge and self-care.

With *The Body Literacy Library* you will learn to tune in to
every little bit of yourself, have all your embarrassing questions
answered, and discover everything you need to know about your
body to live a healthier, happier life. This isn't just about listening
to your body, but empowering yourself with the knowledge
of what your body is telling you.

Read this book to love the skin you're in, and make informed,
positive changes to improve your health and wellbeing.
Starting today.

Contents

-)-)-)-

Introduction

This book is about giving you the understanding you need to look after your skin as well as it can be looked after. Skin occupies a strange place in our lives: on the one hand, many of us – and often society around us – quietly minimize skin problems, put up with years of itch, soreness, or embarrassment, or assume there is nothing much to be done. Access to proper dermatology can be patchy, appointments hard to get to, and it is easy to feel you should not bother anyone.

On the other hand, the very same organ is pulled apart and over-analyzed on social media and in the beauty industry, which can stir up insecurities and sell the idea that the right lotion will erase time, pores, scars, and genes.

Somewhere between the ignored rash and the over-scrutinized pore is where I want this book to sit. A place to see your skin clearly, understand what really matters, and recognize the difference between thoughtful care and wishful thinking in a pretty bottle.

My own way of seeing skin was shaped at St. John's Institute of Dermatology, walking the corridors of one of Europe's busiest skin hospitals and realizing how much of a person's life is written on their surface. The same diagnosis can land entirely differently in two people: one person's psoriasis is a mild nuisance, another's is a full-time job; one person barely notices their rosacea, another rearranges their whole social world around flushing. The story always matters as much as the label.

That is what you see, in a concentrated way, in my TV show, *The Bad Skin Clinic*: the same conditions you read about in textbooks, but lived in real bodies with real fears, families, and futures.

This book will weave between those worlds. We will walk through the anatomy and architecture of the skin; track how it changes from babyhood to old age; unpick the common conditions that fill clinics; and share practical tips and small tricks for the everyday complaints that quietly drive people mad. We will look at how to build a skincare routine that actually suits your skin rather than someone else's, how to read your own patterns, how your skin intersects with other systems in the body, and how to look for proper evidence behind skincare and treatments without needing to become a scientist.

By the time you reach the end of this book, I hope you have a much better understanding of your own skin, that you feel more fond of it, and perhaps even a little bit fascinated by it. If you like, start now: turn your hand over and look at the back, then turn it over again and study your palm. The same organ, two completely different landscapes – fine hairs and freckles on one side, tough, pale, ridged skin on the other. It is extraordinary, and we have not even begun.

So let's get started on the journey of your skin, and help it to stay as healthy as possible, for as long as possible. It has a big job to do.

-)-)-)-

01

Skin deep

skin architecture

Understanding the complexity of the skin requires a closer look at its three fundamental layers – the epidermis, dermis, and hypodermis – as well as the hair, nails, sweat glands, and mucosal membranes that each contribute to the skin's essential roles.

The skin is one of the most remarkable organs in the human body and, fun fact, it's also the largest and the heaviest! It is the boundary between us and the outside world, yet at the same time it is a sophisticated sensory interface, a temperature regulator, and a self-renewing system that never stops working. The skin is, in every sense, the most resilient and multifunctional suit we will ever wear.

The epidermis

The epidermis, the skin's outermost layer, is a powerhouse of regeneration and it performs the crucial function of keeping water in and harmful invaders out, all while being only about 0.1 mm (0.004 in) thick.

The dermis

Beneath the epidermis lies the dermis, the foundation upon which the epidermis rests. The dermis is thicker than the epidermis, varying in depth across different areas of the body. On the eyelids, it is as thin as 0.6 mm (0.024 in), while on the palms and soles, where the skin must endure constant friction and pressure, it can be up to 3 mm (0.12 in) thick. The dermis is divided into two zones:

- **Papillary dermis:** sitting directly beneath the epidermis, this layer is filled with capillaries that provide nutrients to the constantly regenerating epidermal cells. It's also packed with nerve endings that detect touch, temperature, and pain. Fingerprints – our unique ridge patterns – are a direct product of the papillary dermis. These ridges are formed by the dermal papillae, small projections that interlock with the epidermis above, creating friction and enhancing grip.

- **Reticular dermis:** this is where the major collagen bundles are found, acting as a scaffold and ensuring that skin can move without tearing. It is here that we find the sebaceous glands, sweat glands, and hair follicles.

-)-)-)-

THE THREE MAIN LAYERS OF THE SKIN

The skin may look like a single sheet, but it's really three layers. **Epidermis**: our outer shield, this layer renews itself every month; only 0.05 mm (0.002 in) thick on the eyelids but up to 1.5 mm (0.06 in) on the palms and soles. **Dermis**: packed with collagen, elastin, blood vessels, and nerves, this layer provides resilience and sensation; 1–4 mm (0.04–0.16 in) deep. **Hypodermis**: a cushion of fat beneath the top two layers that stores energy and absorbs impact; it varies from a few millimetres to several centimetres deep.

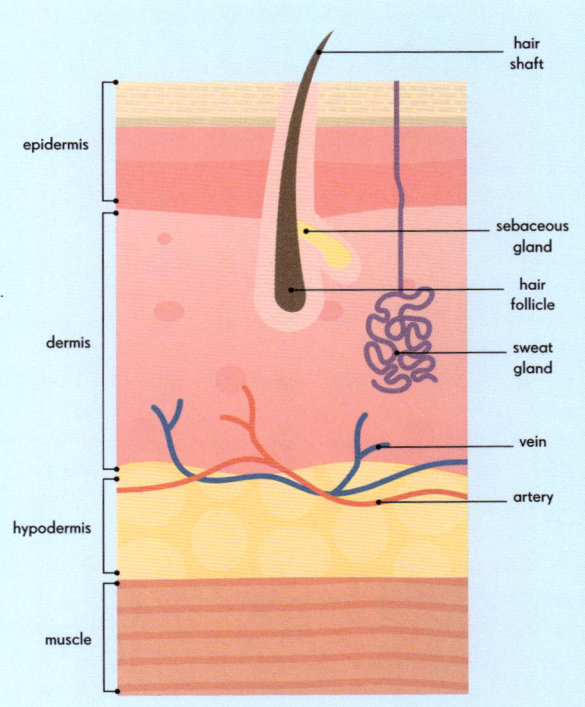

The hypodermis

Moving deeper, we enter the hypodermis, also known as subcutaneous tissue. This layer is composed primarily of adipose (fat) cells, and their role is to provide insulation and shock absorption. Crucially, this is our body's energy store.

The skin is a self-renewing, three-layered suit of protection.

The cells of the epidermis

KERATINOCYTE

The epidermis is made up predominantly of keratinocytes, cells with a singular mission – to form the skin's protective barrier. From the moment a keratinocyte is born, as a plump round cell at the base of the epidermis, it embarks on a roughly 28-day journey upwards through different layers where it is changing shape and performing a different function. Keratinocytes are connected to each other by desmosomes – tiny, rivet-like junctions that provide mechanical strength.

As their name suggests, they begin to produce keratin – the fibrous protein that gives skin its toughness – and lamellar bodies, filled with lipids that create the skin's waterproof barrier.

At the top, in the stratum corneum, the keratinocyte has completed its metamorphosis. It is now a flat, dead cell devoid of a nucleus, packed full of keratin and encased in a layer of fats – thus creating our skin barrier. Here it lingers for a short time before it is naturally shed in a process called desquamation, making way for the next generation of cells rising up from below.

Other cells in the epidermis are:

- **Melanocytes** produce melanin, the pigment responsible for skin colour and UV protection. Exposure to sunlight increases melanin production, leading to tanning or freckling.

- **Langerhans cells** serve as immune sentinels, detecting and neutralizing threats such as bacteria and allergens.

LAYERS AND CELLS WITHIN THE EPIDERMIS

The epidermis is the skin's protective outer layer that contains millions of cells, including protein-forming keratinocytes.

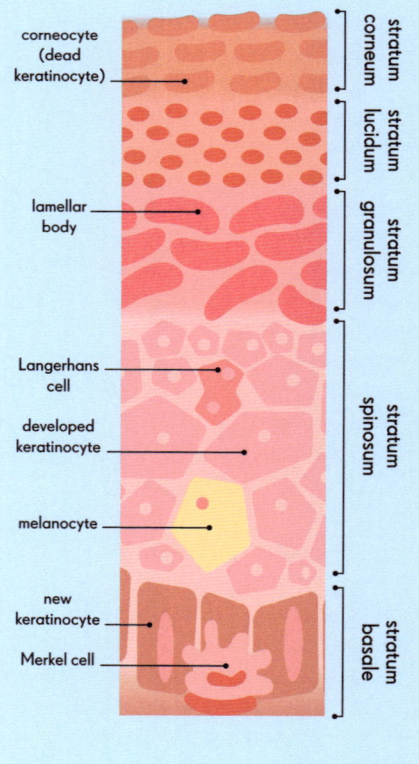

corneocyte (dead keratinocyte) — stratum corneum

stratum lucidum

lamellar body — stratum granulosum

Langerhans cell

developed keratinocyte — stratum spinosum

melanocyte

new keratinocyte — stratum basale

Merkel cell

- **Merkel cells** are concentrated in sensitive areas such as the fingertips, and function as mechanoreceptors, relaying fine touch and pressure sensations to the nervous system.

Hair follicles

We have hair follicles covering most of our body. If you look really hard at your skin you will see very tiny holes, called pores, which are the openings of the hair follicles. Some of the hair is thick and easy to see – called terminal hairs – while other fine, almost invisible hairs are known as vellus hairs.

Hair offers essential sensory functions, protects from cold, UV, dust, and microbes, and also plays a social, cosmetic function.

Each hair follicle is anchored deep in the skin at the hair bulb. Rather like a bulb you would plant in the ground, this is the living centre of the follicle from which the hair shaft grows.

The bulge is the critical zone that houses the hair follicle stem cells. The isthmus, the middle section of the follicle, is where the arrector pili muscle attaches. This small band of smooth muscle responds to temperature changes or emotional stimuli and gives you goosebumps (see pages 38–39).

The infundibulum is located near the top of the follicle. It is a link between the follicle, the skin's surface, and the sebaceous gland, which releases sebum – the oily substance that keeps both the hair and surrounding skin lubricated.

The mucosa

The mucosa is a specialized extension of the skin, lining the body's internal passages, including the mouth, nose, and eyelids. Mucosal surfaces have specialized cells to secrete a lubricating mucus layer as mucosal tissue lacks a structured stratum corneum. This makes it more permeable and susceptible to injury, but the tissues also possess rapid regenerative capabilities.

Nails

Nails grow at an average rate of 3 mm (0.12 in) per month. They are essential for protecting the fingertips and toes, and assisting with tasks requiring fine motor skills.

Tucked beneath the skin at the base of the nail is the nail matrix. This area is responsible for producing keratinocytes, which eventually move forward, become compressed, and form a single strong but flexible sheet that we know as "the nail".

You can often see a whitish, half-moon shape at the base of the nail. This is known as the lunula, and is the visible portion of the active matrix. The nail plate is anchored to the nail bed for nutrients and support.

The cuticle is a thin layer of dead skin that forms a watertight seal to protect the matrix from damage. It always makes me sad that it often gets removed by manicurists. Don't let them – it has a job to do!

-)-)-)-

the skin's immune system

When harmful agents have breached the skin's barrier, it
is the job of our immune system to act as the next layer of
defence. And the skin has its own very adept immune system.

The skin's immune system is highly specialized and composed of two main branches: innate immunity (fast, non-specific responses) and adaptive immunity (targeted, memory-driven responses). Together, these systems deploy an intricate force of specialized immune cells, each with distinct roles in surveillance, attack, and repair.

Innate immunity – the first responders

The innate immune system is fast and always on patrol. Each cell has a particular role to play in protection, and I often think of them like armies with separate, specialized units.

Keratinocytes (see page 14) are the frontline guards, constantly scanning for signs of danger. On their surface sit tiny receptors called toll-like receptors (TLRs), which act as detectors for invaders. When triggered, they sound the alarm and call in reinforcements.

Langerhans cells are the intelligence officers and are constantly sampling what comes in from the outside world. Using long, branching arms, they grab fragments of microbes, then travel to nearby lymph nodes to report in to the adaptive immune system. They're also skilled at telling the difference between harmful invaders and harmless bystanders, which helps prevent unnecessary inflammation.

Macrophages are the skin's clean-up crew and heavy-lifters. They hang out deeper in the dermis, looking for damaged cells or invaders to engulf and digest. Once they've dealt with a microbe, they display a fragment of it on their surface like a wanted poster, priming the adaptive immune system for future encounters.

Neutrophils are the emergency response team. They travel in the bloodstream and are the first to arrive at any sign of infection. Once on site, they either engulf the invader or release toxic granules to destroy it. If things get really dire, they can self-destruct, releasing sticky webs of DNA and enzymes (called NETs) to trap and kill microbes, and alert other immune cells to join the fight.

Natural killer (NK) cells are the body's undercover assassins. They constantly scan for cells that look suspicious – like those infected by viruses or turning cancerous – and destroy them. They have two weapons: perforin (which punches holes in the target's membrane) and a death signal (which tells the cell to self-destruct safely).

Mast cells are the signal flares. When they detect trouble, they release histamine and other chemical messengers that make blood vessels wider and leakier. This allows immune cells to flood into the area quickly – like lifting a drawbridge to let reinforcements pour in.

Eosinophils are the big-game hunters. Their speciality is tackling parasites, especially large ones like worms that other cells can't swallow. They're summoned into the skin during allergic reactions or chronic inflammation, where they release toxic granules to break down their oversized targets.

Adaptive immunity – the memory keepers

Unlike the urgency required in the innate immune system, the adaptive immune response is precise and responsible for developing long-term immunity. This branch of the immune system relies on lymphocytes (white blood cells) known as T cells and B cells, each designed to recognize and fight a particular invader. Two types of T cells play critical roles:

01. **Cytotoxic T cells** (CD8+ T cells): these are responsible for destroying virus-infected and cancerous cells by directly binding to them and initiating cell death.

02. **Helper T cells** (CD4+ T cells): these coordinate the immune response by releasing signalling molecules (cytokines) that activate other immune cells, including macrophages and B cells.

B cells are responsible for long-term immune memory, producing specific antibodies (immunoglobulins) that target and neutralize pathogens. Once a B cell encounters its target antigen, it differentiates into plasma cells, which mass-produce antibodies that help destroy the invader. B cells also generate memory B cells, which remain in the body for years. If the same pathogen is encountered again, these memory B cells mount a rapid and targeted response, ensuring faster clearance of the infection.

The innate immune system is a fast-acting mobile defence force that reacts to threats within the body.

-)-)-)-

protection

Your skin is not just a covering — it is a highly sophisticated defence system, working around the clock to repel harmful invaders while holding onto essential elements, such as water and nutrients.

I have likened the skin to a suit of armour, and that's exactly what it is. It's hard to believe the number of defence systems this organ has in place to protect the rest of our body. Even the smallest breach in this armour can have a significant impact on our health. I'll explore this in later chapters, but first, let's consider the many ways our skin defends us.

The skin is the body's first line of defence, acting as a physical barrier, a repair system, a biochemical shield, and even a built-in sunscreen. It protects us from external threats like pathogens, trauma, and environmental changes, while also regulating our internal environment to safeguard the biological functions that keep us alive and well.

The skin barrier

The stratum corneum, the outermost layer of the epidermis, is an extraordinary biological shield that prevents water loss and blocks harmful substances from entering. Often described as a "bricks-and-mortar" structure, this skin barrier consists of corneocytes (flattened, dead skin cells) held together by corneodesmosomes (sticky structures) within a lipid matrix composed of ceramides,

cholesterol, and free fatty acids. This carefully arranged structure makes the skin waterproof, resilient, and able to withstand daily environmental stress.

Each corneocyte is packed with keratin, a fibrous protein that provides mechanical strength. It is encased in a tough (cornified) envelope made of cross-linked protein-derived natural moisturizing factors (NMFs). These elements help the skin retain moisture and remain flexible. The lipid matrix between the corneocytes forms a hydrophobic barrier, preventing excessive transepidermal water loss (TEWL) and stopping microbes, allergens, and irritants from penetrating.

This defensive layer is not static – it is constantly shedding and renewing itself, ensuring that damage is quickly repaired.

The dangers of no protective layer

The absence of this barrier and its waterproofing function would lead to uncontrollable water balance, meaning we would either dissolve or totally swell up. During a simple rain shower, our bodies would rapidly absorb water, causing swelling and allowing bacteria, fungi, and viruses to enter the skin, potentially causing harm.

-)-)-)-

THE "BRICKS-AND-MORTAR" COMPOSITION OF THE STRATUM CORNEUM

The stratum corneum comprises large, flat, polyhedral cells filled with keratin.
These cells are arranged in vertical layers and connected by a lipid-rich cement,
similar to a construction of "bricks and mortar".

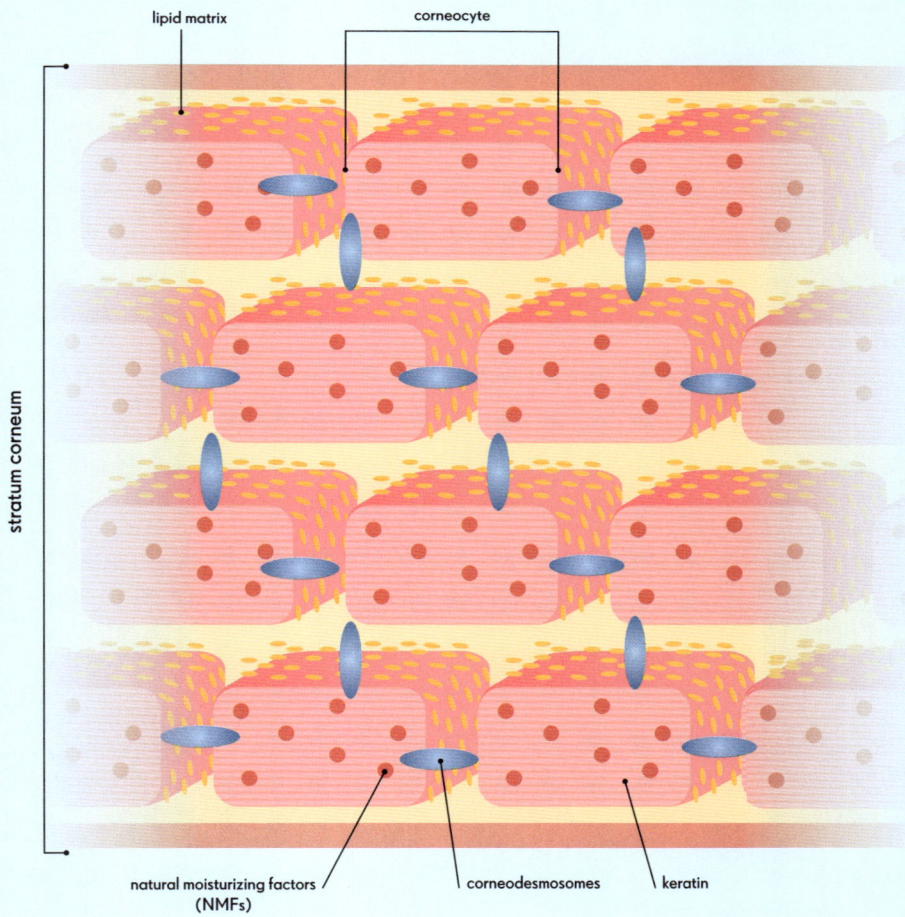

lipid matrix

corneocyte

stratum corneum

natural moisturizing factors
(NMFs)

corneodesmosomes

keratin

The acid pH mantle

Beyond its structural strength, the skin has a chemical defence system known as the acid mantle – an invisible, biochemical shield that maintains the skin's pH between 4.5 and 5.5.

This acidic environment is generated by several processes, including:

01. The fermentation of sebum and sweat by bacteria, producing weak organic acids such as propionic, acetic, and lactic acid.
02. The breakdown of structural proteins like filaggrin into natural moisturizing factors.
03. The release of free fatty acids, such as oleic and linoleic acid, from sebum.
04. The metabolism of ceramides and other lipids.

When this delicate pH range is breached, often due to the use of harsh chemicals, skin conditions such as dermatitis can arise. People with atopic dermatitis often have a higher skin pH (around 6.5), which encourages bacterial overgrowth and increases the risk of infection. Similarly, ageing skin and chronic wounds tend to become more alkaline, leading to poor wound healing and greater susceptibility to disease.

Protection against UV radiation

Although a certain amount of ultraviolet (UV) radiation is good for us, we also recognize that it is harmful to our skin – it damages the DNA inside our skin cells and unleashes a cascade of reactive molecules known as free radicals. This process is called oxidative stress and it harms the skin cells. The skin has its own natural antioxidant system that is constantly working to neutralize these free radicals, but if there are too many of them then damage is done.

To try and protect the cells from direct DNA damage from UV radiation, the skin relies on melanin – a natural pigment made by cells called melanocytes. All skin colours have about the same number of melanocytes, but the type and quantity of melanin they produce differ. Eumelanin (a dark brown or black pigment) offers better protection, while pheomelanin (a red or yellow pigment, more common in people with fair skin and red hair) is less effective at absorbing UV.

Melanin production

When your skin is exposed to sunlight, it switches on an enzyme called tyrosinase. This acts like a chemical spark, triggering a chain reaction. Once active, tyrosinase takes a natural ingredient already in the skin (the amino acid, tyrosine) and turns it into the raw material for melanin. The melanin is then packed into tiny parcels called melanosomes, which are then passed from the pigment-making melanocyte into the surrounding skin cells. These melanosomes sit over the nuclei of your keratinocytes, shielding the DNA from UV rays like a microscopic parasol (see image right).

-)-)-)-

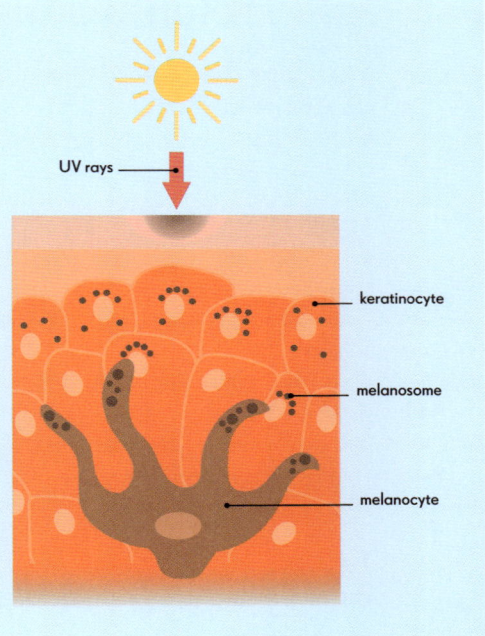

BEHIND THE SUN TAN

The movement of melanin pigment in the melanosome is essential to protect the keratinocytes when there is UV exposure.

UV rays

keratinocyte

melanosome

melanocyte

The hypodermis

Beneath the dermis, the hypodermis (or subcutaneous fat layer) plays a crucial role in thermal insulation, cushioning, and energy storage. This layer of fat, or adipose tissue, acts as a shock absorber, protecting bones and muscles from injury. The thickest deposits of subcutaneous fat are found in high-impact areas, such as the buttocks, thighs, and abdomen, providing extra padding against falls and collisions. Specialized fat pads in the soles of the feet and palms absorb pressure, reducing strain on joints.

In cold environments, this layer traps heat, preventing rapid temperature loss. Adipose tissue is a poor conductor of heat, so it slows down the transfer of warmth from the body to the environment. This is why individuals with a thicker subcutaneous fat layer tend to have greater resistance to the cold.

-)-)-)-

temperature control

No matter the season, whether you're bracing against an icy wind or basking in summer heat, your body is constantly fine-tuning its internal temperature. Let's take a look at the role that the skin plays in this process.

Every second, without you realizing, your skin makes micro-adjustments – opening blood vessels, producing sweat, and raising tiny hairs – to ensure that your core temperature stays within its narrow safe zone of 36–38°C (96.8–100.4°F). Drop below this, and hypothermia sets in; go too high, and the enzymes that power every cell in your body start to break down.

The balancing act

Deep in the brain, the hypothalamus – your body's internal thermostat – monitors temperature fluctuations, detecting signals from thermoreceptors in the skin and bloodstream. The skin, as the body's largest organ, is the first responder, deploying an intricate system of sweat glands, blood vessels, hair follicles, and fat stores to maintain thermal balance.

When we are too hot

SWEATING

When the mercury rises, your body's most powerful cooling system kicks in: sweating. Coiled deep in the dermis, eccrine sweat glands spring to life, releasing moisture onto the skin's surface. But sweating alone doesn't cool you – evaporation does. As sweat transforms from liquid to vapour, it pulls heat away from the skin, cooling both the surface and the blood circulating beneath it. This cooled blood then travels back to the body's core, ensuring that vital organs remain within their safe operating temperature.

Humans are unparalleled when it comes to endurance cooling. Unlike panting dogs or fur-covered mammals that must pause to regulate their temperature, we can shed heat continuously. Our sweat glands are so efficient that, under extreme conditions, some people can sweat up to 3 litres (5.3 pints) per hour – a rate unmatched by any other species. This ability gave our ancestors a survival advantage, allowing them to outrun prey in endurance hunts through a phenomenon known as persistence hunting.

VASODILATION

Sweat isn't the only cooling trick up the skin's sleeve. As temperatures climb, blood vessels in the dermis widen (vasodilation), allowing warm blood to flow closer to the surface, where heat is radiated outwards. If a warm breeze passes over

-)-)-)-

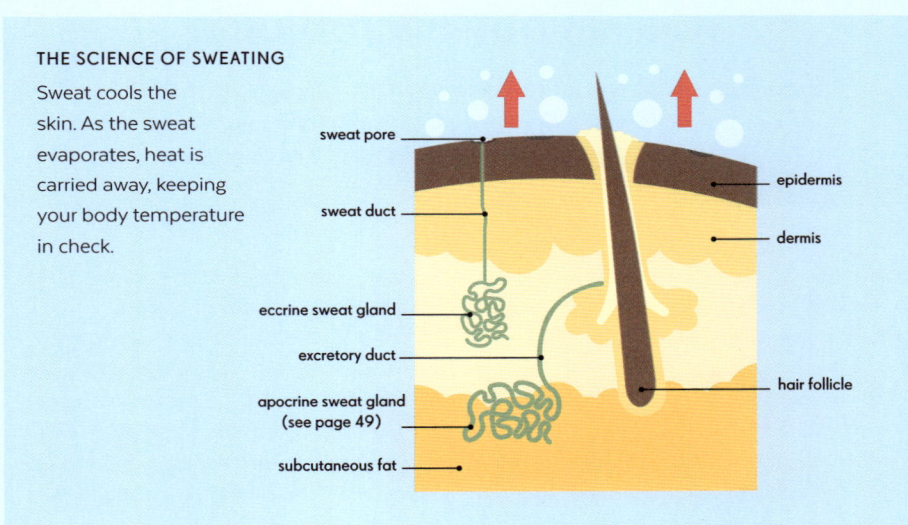

THE SCIENCE OF SWEATING

Sweat cools the skin. As the sweat evaporates, heat is carried away, keeping your body temperature in check.

sweat pore

epidermis

sweat duct

dermis

eccrine sweat gland

excretory duct

apocrine sweat gland
(see page 49)

hair follicle

subcutaneous fat

your skin or you press a cool glass against your wrist, the heat transfer is almost immediate. Radiation, convection, and conduction – the three main ways that heat escapes – work together to keep your body from overheating.

When the chill sets in

What about when the temperature plummets? When your skin signals that heat is being lost too quickly, the hypothalamus shifts into heat-retention mode, activating three key responses:

01. Vasoconstriction: Blood vessels in the skin narrow, reducing circulation to the surface. This keeps warm blood concentrated in the body's core, preventing heat loss from exposed areas.

02. Piloerection (goosebumps): The tiny arrector pili muscles at the base of each hair follicle contract, making hairs stand upright. In our fur-covered ancestors, this created an insulating layer of trapped warm air; today, it's mostly an evolutionary relic, but still a sign of your body's effort to retain heat.

03. Shivering: When temperatures drop too low, the body generates heat through involuntary muscle contractions. These tiny, rapid movements burn energy, releasing heat as a byproduct. Infants, who lack the muscle mass for effective shivering, rely instead on brown adipose tissue (BAT) – a special type of fat that metabolizes heat rather than storing energy.

-)-)-)-
·.·.·.

your unique microbiome

Every inch of our skin is teeming with life. Though we may think
of our bodies as our own, we are in fact walking ecosystems,
hosting an invisible but essential world of microbes.

Within the skin, microbial cells vastly outnumber human cells. The estimated ratio of microbial cells to human skin cells is roughly 10 to 1.

What is the skin microbiome?

The skin microbiome is a vast community of bacteria, fungi, viruses, and even microscopic mites that inhabit our skin. The resident microbes in our skin are not passive bystanders; instead, they contribute actively to maintaining the skin barrier, supporting immune function, and even aiding in wound healing. By producing specialized molecules and metabolites, these microbes communicate with skin cells, regulating inflammation and reinforcing the skin's defences.

REGIONS OF THE SKIN MICROBIOME
Your skin varies across different body regions, featuring dry, moist, and oily zones. Each of these areas hosts a unique microbiota that has adapted to thrive in its specific environment.

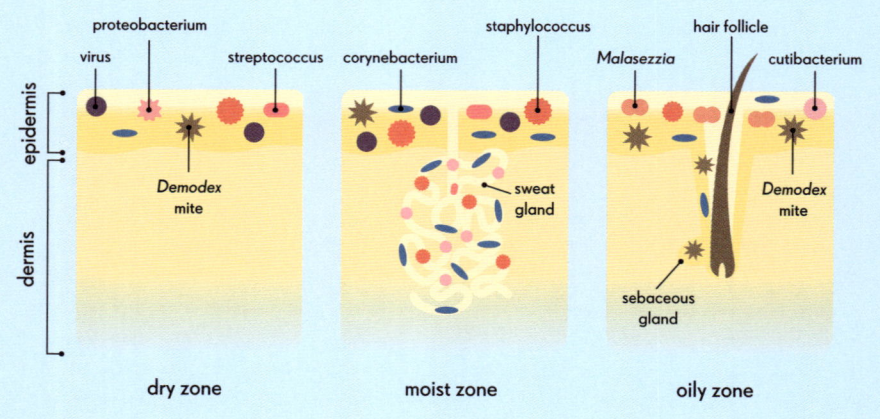

dry zone   moist zone   oily zone

Different parts of our skin each have their own unique microbial make-up. This is why some skin conditions, like eczema, often appear in specific areas, such as the inside of the elbows, while others, like psoriasis, show up on the outer elbows. Interestingly, no two people have the exact same skin microbiome. The composition of microbes on your skin is shaped by your genetics, age, diet, lifestyle, environment, and even the people you live with.

Say hello to your little friends

Far from being an uninvited nuisance, the microbes that live on our skin are integral to its function and protection. The bacterial population is dominated by three major groups: staphylococci, corynebacteria, and cutibacteria. These microbes are found in different densities across the body, adapted to thrive in distinct environments – oily, moist, or dry. For example, *Cutibacterium* acnes prefer the oil-rich environment of our hair follicles, where it helps break down sebum but, when overactive, can contribute to acne. In dry environments, the microbiome favours hardy bacterial communities that can tolerate low humidity and a disrupted barrier. In these areas, proteobacteria often become more dominant, while *Streptococcus* species are less prevalent.

Staphylococcus epidermidis, a usually friendly resident, helps prevent harmful bacteria from invading by competing for resources and producing antimicrobial substances. Yet even seemingly helpful microbes can turn against us under the right (or rather, wrong) conditions – *Staphylococcus aureus*, which many people carry harmlessly, is also responsible for serious skin infections when it gets out of control.

Another area of interest is how our environment shapes the microbiome. People who grow up in urban areas, where there is less exposure to diverse microbes, tend to have less microbial diversity on their skin compared to those who spend more time in nature. Some studies suggest that early exposure to a variety of microbes may help train the immune system and reduce the risk of allergic conditions such as eczema.

• PROTECTING THE MICROBIOME •

The optimal approach to safeguarding your skin microbiome involves gentle cleansing, avoiding antibacterial products unless absolutely necessary, prioritizing adequate sleep, and adopting a colourful, plant-rich, wholefood diet.

-)-)-)-

Skin microorganisms

As well as bacteria, we have an array of tiny microorganisms that live on our skin, not just coming along for the ride but active participants in the health of our entire organism. I always imagine that our body must be like a giant planet to them. Among the most intriguing inhabitants of human skin are *Demodex* mites – tiny, translucent arachnids that live in our hair follicles. There are two primary species: *Demodex folliculorum* and *Demodex brevis*.

The former prefers to reside in hair follicles, particularly around the face, while *D. brevis* burrows deeper, residing within sebaceous (oil) glands.

Demodex mites are nocturnal, emerging at night to move across the skin and mate before retreating into follicles by day. Most people host them without issue, but when their population grows unchecked – often due to immune imbalances or changes in skin oil production – they can contribute to skin conditions such as rosacea and blepharitis.

ANATOMY OF A DEMODEX MITE

As unpleasant as *Demodex* mites might look up close, we all have them in our skin microbiome, especially in areas of the face such as the hair follicles of our eyelashes.

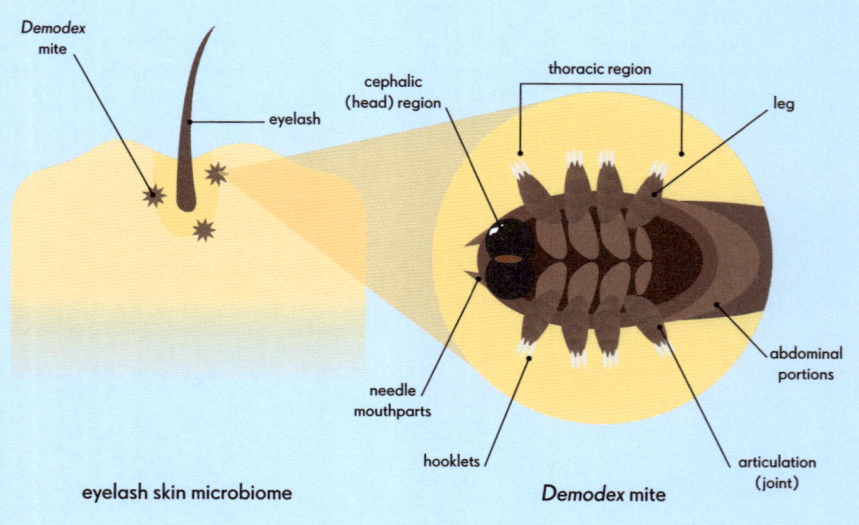

Demodex mite

eyelash

cephalic (head) region

thoracic region

leg

needle mouthparts

hooklets

abdominal portions

articulation (joint)

eyelash skin microbiome

Demodex mite

The role of viruses and yeasts

While bacteria and fungi get most of the attention, the skin is also home to a variety of viruses, including bacteriophages – viruses that infect bacteria. These phages help regulate bacterial populations, ensuring that no single species dominates. Some scientists are even investigating ways to harness bacteriophages as a natural means of controlling antibiotic-resistant bacteria.

Another major player in the skin microbiome is *Malassezia*, a genus of yeast that thrives on our skin's natural oils. Unlike harmful fungi that cause infections, *Malassezia* typically exists as a harmless organism. However, under certain conditions – such as increased humidity, excessive oil production, or weakened immune defences – these yeasts can shift from quiet residents to troublemakers.

Overgrowth of *Malassezia* has been linked to conditions like seborrhoeic dermatitis (dandruff), pityriasis versicolor (a skin pigmentation disorder), and even some forms of eczema. Interestingly, despite its ability to cause irritation, *Malassezia* also appears to play a role in regulating the skin's microbial balance. The key, as always, lies in maintaining equilibrium.

The study of the skin microbiome is evolving rapidly, and scientists are now exploring ways to manipulate it to improve skin health. This could involve probiotics (introducing beneficial bacteria), prebiotics (providing nutrients to support good microbes), or even bacteriophage therapy (using viruses that target specific bacteria).

It is clear that the tiny microbes, organisms, and viruses on our skin are far more important than we once thought.

-)-)-)-

DOES SWEATING "DETOX" ME?

If you think sweating it out in a sauna is going to detox you, it doesn't quite work like that. The real detoxification work in your body is done by the liver and the kidneys, and they process and eliminate metabolic by-products, drugs, environmental chemicals, and more. Some studies have detected certain toxins in sweat, but their amounts are extremely low. The sweat glands don't have active detox mechanisms – they are just passive tubes. Sweat's main purpose is temperature regulation, and so it contains electrolytes, lactate, and urea, but not other elements that you might think of such as alcohol or bad food! What sweating can do, however, is add support to the skin's microbiome.

CAN I SHRINK MY PORES?

A pore is actually the opening of the hair follicle unit with its attached oil gland, so you cannot meaningfully shrink the anatomical opening. You can, though, make the pores "look" smaller and tighter by managing what makes them look bigger. What actually enlarges a pore is a combination of three things. First, excess oil from the glands loosens the follicular wall, making the opening look wider. Second, micro-comedones – those tiny plugs of keratin and oil that sit just beneath the surface – push the walls apart. And third, loss of dermal support: when collagen declines with age or sun damage, the pore rim slackens, making it appear larger. So, what you can do is: 1. Regulate oil production – topical retinoids are the gold standard. 2. Improve the smoothness of the pore wall – chemical exfoliants such as salicylic acid (a beta-hydroxy acid) can dissolve the oil and keratin inside the pore, so the opening sits flatter and reflects light more evenly. 3. Support the dermal architecture – sun protection is non-negotiable to protect against UVA, which weakens collagen around the pore and exaggerates laxity.

ARE WHITE SPOTS ON MY NAILS A SIGN OF CALCIUM DEFICIENCY?

Small, chalky flecks on the nails are a sign of trauma to the nail matrix rather than a nutritional deficiency. The nail matrix is the tiny factory under the cuticle that produces new nail. If you knock it (even lightly, often without noticing – typing, picking, manicures), it can momentarily disrupt keratin formation and leave a pale, opaque spot as the nail grows out. Severe deficiencies in minerals or proteins can affect overall nail quality, but that is rare and looks very different – brittle, ridged, slow-growing nails rather than isolated white specks.

—

ARE GEL NAILS SAFE?

Gel UV (ultraviolet) manicures are not "toxic", but they are not completely benign either. The main issues are UV exposure from the lamps, allergy to the (meth)acrylate that is used, and the way that the nails are applied and removed. Gel nails are a mix of (meth)acrylate compounds, a photo-initiator (which makes the gel harden under UV light), pigments, and fillers. Under UVA light from the lamp, the photo-initiator breaks apart and starts a chain reaction that connects the small acrylate units into a hard, cross-linked plastic coating stuck to the nail plate. The problems arise from the light needed to cure the gel nail and from any uncured compound that contacts the skin. UV lamps emit predominantly UVA light, which penetrates to the dermis and is linked to photo-ageing and skin cancer. UV nail lamps are classified as low-risk devices when used according to instructions, but sensible precautions should be taken such as sunscreen or protective gloves. (Meth)acrylates can increase the risk of allergy – once sensitized, patients may react to acrylates used in dental composites, some medical devices, and other adhesives. Photo-initiators may be more of a problem, especially one called TPO – recent toxicology work in animals reclassified this as a suspected carcinogenic. The mechanical side, however, is what actually ruins nails – removing the cuticle, over-buffing the plate, and prolonged use of acrylates.

—

02
The great communicator

touch

The first of our senses to develop in the womb,
touch is a silent language between mother and
child before any other connection is possible.

Throughout life, touch remains fundamental, but for all its gentleness, it's also the sense through which we experience pain – perhaps our most primal and urgent warning system. A body without pain is a body without protection. The ability to sense pressure, temperature, and injury has been refined over millions of years, ensuring that we respond appropriately to threats.

Skin cells involved in touch

Most critical for touch and feel is, of course, the skin on the palms of our hands and fingertips and on the soles of our feet. These areas feature a detection system of four types of specialized cells that we collectively call mechanoreceptors.

Merkel cells are tiny, round-shaped cells found at the base of the epidermis. They detect fine details and delicate textures, like the tiny ridges on a grain of rice.

Meissner corpuscles are dome-shaped cells that sit just beneath the surface of the skin. These are responsible for detecting subtle shifts in an object's position. They are also the reason we feel the sensation of fabric when we first get dressed, but they quickly adapt and stop responding.

Deeper in the skin are Pacinian corpuscles that detect high-frequency vibrations – like the hum of an electric toothbrush – and help us assess movement within objects we are handling.

Finally, there are Ruffini endings, which are responsible for detecting skin stretch and joint position. These receptors enable us to maintain a sense of body awareness even without looking, ensuring that we know where our fingers are in space as we type on a keyboard or tie a shoelace.

The need for touch extends way beyond the infant years into adulthood.

From the skin to the brain

Mechanoreceptor signals travel from the skin to the brain via A-beta fibres – large, myelinated (sheathed) nerves that act like high-speed

THE HOMUNCULUS

The homunculus isn't a literal structure but a representation to show how much brain area is devoted to sensing different body parts. Scaled by touch, it gives us enormous tongues and fingers compared to our legs and arms.

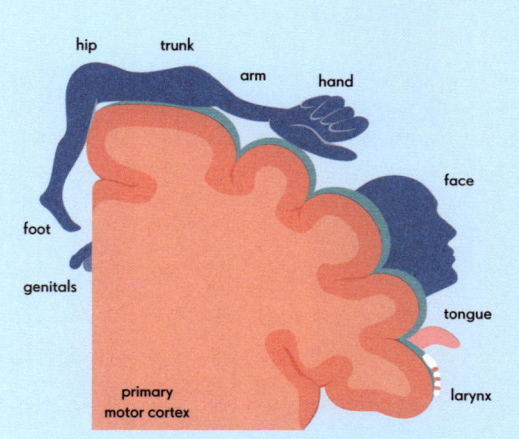

hip   trunk   arm   hand   face   foot   genitals   tongue   larynx

primary motor cortex

broadband, transmitting precise touch, pressure, and vibration at up to 75 metres (246 ft) per second. The brain relies on this fast information to react quickly – whether to adjust grip, avoid harm, or navigate fine textures. These signals are delivered to the somatosensory cortex, the brain's touch-processing centre, where they're mapped in a detailed sensory layout known as the homunculus. This quirky, distorted "body map" reflects how much brain space is devoted to each area – so hands, lips, and face appear far larger than the rest.

While other touch receptors quickly relay sharp details about pressure and texture, C-tactile fibres are tuned for emotional touch and they take their time. Found in the hairy skin of the body, these nerves respond to things like a gentle stroke or a comforting pat. Instead of rushing the message to the brain, they send it slowly, giving the brain time to process not just the touch, but how it feels. Their signals go to the brain's emotional centres, where feelings and memories live, making these fibres key to the soothing power of human connection.

The importance of this system can be seen in the comforting effect of a gentle touch, whether it be a reassuring pat on the back or the warmth of a parent's hand. Studies have shown that touch plays a crucial role in early development, with premature babies who receive regular skin-to-skin contact (often known as "kangaroo care") displaying better overall health outcomes. In adults, touch triggers the release of oxytocin, sometimes called the "love hormone", which strengthens social bonds, and lowers levels of the stress hormone cortisol.

Interestingly, context influences how touch is perceived. A light touch from a loved one feels entirely different from the same touch delivered by a stranger or in an unexpected situation!

The skin's pain receptors

Pain is the body's oldest alarm system. It is governed by specialized nerve endings called nociceptors, which lie like the fine, tangled roots of a plant, waiting for trouble. They remain silent through everyday sensation, only sparking to life in the presence of real threat. Once triggered, they send signals racing along dedicated fibres to the spinal cord, and from there, to the brain – where pain becomes something we feel, interpret, and, ideally, act on.

HOW NOCICEPTORS RELAY PAIN TO THE BRAIN

That sharp prick? Nociceptors turn it into lightning-fast electrical impulses, shooting along nerves and spine to the brain, which translates it into pain.

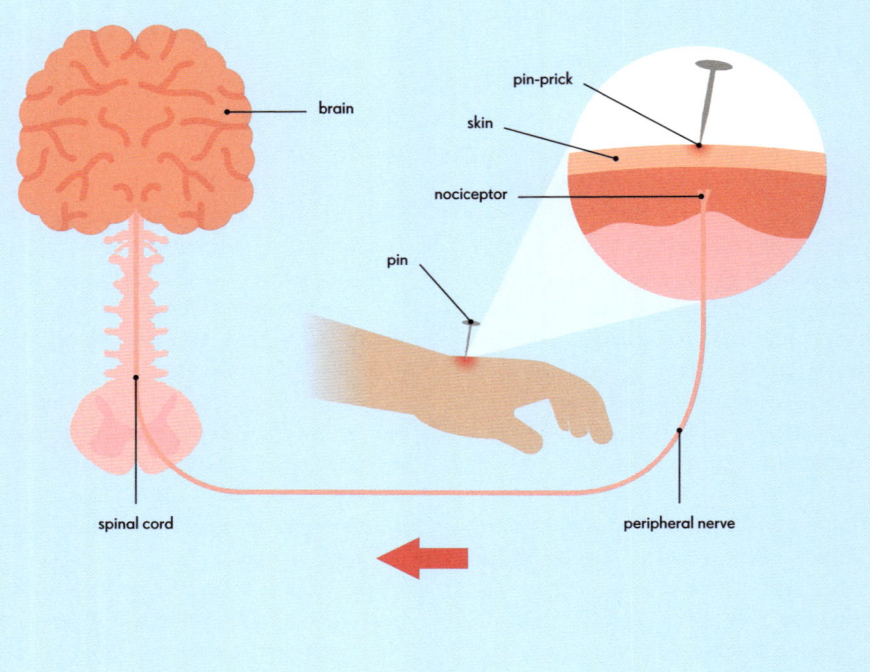

brain

pin-prick

skin

nociceptor

pin

spinal cord

peripheral nerve

-)-)-)-

Pain processes

Pain acts in two ways. The first is sharp and immediate – think about the jolt that pulls your hand away from a hot pan before you've had time to think. This is the work of A-delta fibres, thin, myelinated nerves that conduct signals at high speed to the spinal cord and brain at speeds of up to 35 m/s (115 ft/s). Their message is urgent: act now, protect yourself.

The second wave is delivered more slowly as a dull, lingering ache to remind you not to use that hand too much for the rest of the day. This is carried by C-fibres, unmyelinated and far slower at about 2 m/s (6.6 ft/s). They make sure we don't forget the injury too quickly and allow it to heal.

Pain is meant to protect us, but sometimes that alarm misfires. Allodynia is when normally harmless sensations, like a sleeve lightly brushing sunburnt skin, feel painful. This happens when inflammation lowers the threshold of pain receptors, making them hypersensitive. While it's meant to prevent further injury, in some people the nerves don't reset. Instead, the system becomes stuck and the brain and spinal cord keep amplifying pain signals, even after the injury is gone. The result is chronic pain that lingers without cause.

The mind's role in pain

Pain is not just a physical event – it is also shaped by the mind. Psychological factors, including fear, anxiety, and expectation, can heighten or dull pain perception. Fear and anticipation can intensify pain. For example, the sight of a needle before an injection often causes more distress than the injection itself. Stress hormones, like cortisol, can amplify pain sensitivity, making what should be a minor discomfort feel far worse. This highlights the complex interplay between the body and the mind in shaping our experience of pain.

• IMPERVIOUS TO PAIN •

Some people feel no pain at all - they are born with a rare condition called congenital insensitivity to pain. They can leap from heights, walk across coals, or pierce their skin without flinching. But bones still break, organs bruise, and wounds infect. Without pain as a warning, serious harm is often the consequence.

why we itch

Itch is one of the body's most primal and instinctive
responses. In its simplest form, the sensation of itch prompts
an immediate reaction – scratching – to remove potential
dangers from the skin before they can cause harm.

Itch has its own dedicated nerve fibres and
signalling pathways, separate from pain. Unlike
pain, which provokes withdrawal and avoidance,
itch triggers an entirely different response:
scratching. This distinction suggests that itch
evolved to target specific threats, particularly
tiny invaders such as parasites, rather than the
larger-scale dangers that induce pain. From insects
burrowing into our flesh to toxic plants brushing
against our skin, the itch response is a rapid and
effective way to trigger protective behaviour.

> ## • CONTAGIOUS ITCHING •
>
> An intriguing aspect of the itch
> response is contagious itching.
> Simply seeing or thinking about
> itching can trigger the sensation –
> perhaps you are scratching a part of
> your body or maybe the top of your
> head has suddenly started feeling
> very itchy while you read this!

What exactly is an itch?

At the core of the itch response are specialized
nerve fibres in the skin called pruriceptors. These
fibres detect itch-inducing stimuli and send
signals through the spinal cord to the brain.
Unlike pain fibres, which conduct signals rapidly
to trigger an immediate response, itch signals
move far more slowly – at just 2 m (6.6 ft) per
second, compared to the 80 m (262 ft) per
second of pain. This sluggish transmission may
be why itch tends to build gradually, creating
an unbearable urge that demands relief.

One of the best-known triggers of itch is
histamine, a molecule released by mast cells in

response to allergens, insect bites, and irritants.
Histamine binds to receptors on pruriceptors,
activating them and sending an itch signal to
the brain. This explains why antihistamines are
often used to treat allergic itch, as they block
histamine receptors and dampen the response.
It's important to know that histamine is just one
piece of the puzzle and many forms of itch do
not respond to antihistamines, indicating that
other pathways are at play.

Another molecule called brain natriuretic
peptide (BNP) plays a key role in transmitting itch
signals without triggering pain. BNP is released
from certain nerve cells in response to itch stimuli,

-)-)-)-

THE ITCH–SCRATCH CYCLE AND HISTAMINE RELEASE

Occasionally, the act of scratching can be so pleasurable that it triggers a repeated behaviour known as the itch-scratch cycle.

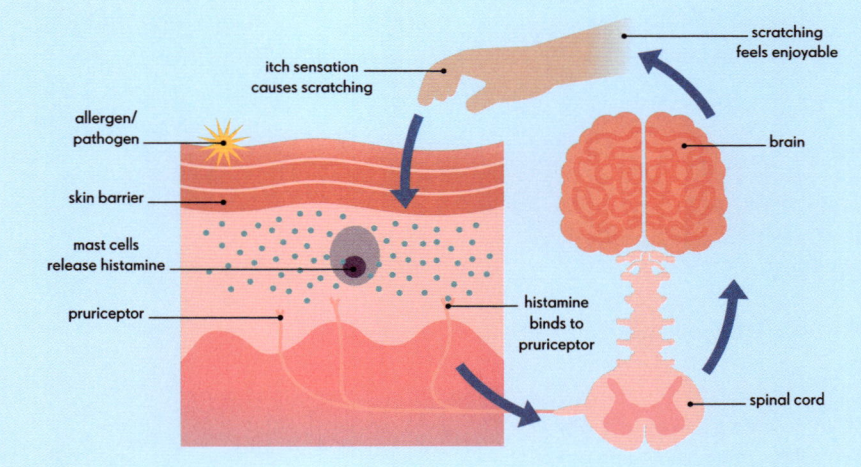

activating a separate pathway that travels to the brain's sensory processing centres. The spinal cord acts as a processing hub for itch. A specific group of inhibitory spinal interneurons function as gatekeepers, deciding whether an itch signal reaches the brain. When these neurons are active, they suppress itch signals, preventing unnecessary scratching.

Other itch stimuli

Beyond chemical triggers, mechanical stimuli such as a light touch on hairy skin can also induce itch. This sensitivity to light touch is thought to be an evolutionary adaptation, helping us detect and remove crawling insects before they bite or burrow. However, the nervous system has an inbuilt paradox: scratching an itch provides relief because pain signals temporarily override the itch sensation. This is why we instinctively scratch mosquito bites or rashes – brief pain suppresses the itch. Unfortunately, this relief is short-lived, and excessive scratching can cause further inflammation and even infection, creating a vicious itch-scratch cycle.

The skin microbiome (see pages 24–27) also plays a role as certain microbes can trigger itch directly. For instance, *Staphylococcus aureus*, a bacterium commonly found on the skin, has been shown to release a protein that directly activates nerve cells, causing an intense itch.

-)-)-)-

emotional signals

The skin's emotional responses are often beyond our conscious control. The fine hairs on our arms rise as goosebumps when something wonderful happens, while the deep flush of a blush arrives uninvited in moments of self-consciousness.

Goosebumps and blushing are responses that are designed to signal our emotions to those around us and help us navigate the social world.

But how does this happen? The skin is embedded with a sophisticated network of sensory neurons, blood vessels, and sweat glands, all tightly linked to the autonomic nervous system. This means that emotions – originating in the brain – can ripple through our skin before we even have a chance to process them consciously.

Goosebumps

Goosebumps are a fascinating leftover from our evolutionary past, a physical reaction deeply wired into our nervous system. The technical term for this response is piloerection, and it occurs when tiny muscles at the base of each hair follicle – arrector pili muscles – contract, causing the hairs on our skin to stand on end. We get goosebumps whenever we experience something thrilling or moving due to a surge of activity in the sympathetic nervous system that our brain interprets as a moment of heightened significance. In response, the amygdala, the part of the brain responsible for

processing emotions, signals the release of adrenaline. This hormone triggers a cascade of physiological effects, including increased heart rate, dilated pupils, and, crucially, the activation of the arrector pili muscles.

Blushing

Another emotional signal displayed by the skin is blushing, which is a uniquely human phenomenon – an involuntary, physiological response that visibly broadcasts our emotions to the world. Here there is another burst of activity in the sympathetic nervous system that releases adrenaline. One of the main effects of this is vasodilation, or the widening of blood vessels. While this happens throughout the body, the effect is particularly pronounced in the facial blood vessels. Unlike other blood vessels, which can contract to limit the effects of vasodilation, those in the face lack this regulatory control. As a result, more blood rushes to the surface of the skin, creating the unmistakable flush of colour we recognize as blushing.

Blushing also has a strong connection to romantic attraction. The same surge of adrenaline that fuels a racing heart and sweaty

-)-)-)-

THE SCIENCE OF GOOSEBUMPS

Goosebumps arise when small arrector pili muscles within the skin contract, causing a slight elevation of the skin and hair. This predominantly occurs on the arms and legs.

skin

goosebump

hair   arrector pili
        muscle relaxed

arrector pili
muscle contracted

palms can also cause the cheeks to flush, serving as an unintentional but obvious sign of interest. This effect is so deeply ingrained in human psychology that some cultures have embraced it, with blushing cheeks being considered a sign of beauty and youth – hence the long-standing tradition of using blush or rouge in make-up.

Studies have also shown that people who blush after a social misstep are often perceived as more trustworthy and likeable, reinforcing the idea that this reaction may help maintain social bonds.

The skin's changes reveal our inner emotional state, whether we want them to or not.

facial expressions

A fundamental way that we communicate emotion is through
facial expressions, and the skin plays a central role in this process.

The face is unique in that it has a high concentration of small, delicate muscles that insert directly into the skin. This allows for an extraordinary range of movement, from the slightest twitch of an eyebrow to a broad, expressive smile. These movements are not only controlled by the brain but are also influenced by sensory feedback from the skin itself, which helps fine-tune expressions and make them more precise.

movement subtly shifts the overlying surface, creating the visible expressions that others interpret.

Facial expressions are both voluntary and involuntary. We can control some movements deliberately, such as forcing a smile in a photograph, but many expressions occur automatically in response to emotion, sometimes before we are consciously aware of them. It is useful to be aware of the resting expression of your face!

Smiles, frowns, and furrowed brows

The human face contains around 20 muscle pairs dedicated to facial expression. We have coded around 10,000 different facial expressions associated with emotion and all are controlled by the facial nerve (cranial nerve VII). When we smile, the zygomaticus major muscle contracts, pulling the skin at the corners of the mouth upwards.

A frown occurs when the corrugator supercilii muscle pulls the skin of the forehead and eyebrows together, creating visible furrows. Expressions of surprise involve the frontalis muscle lifting the skin of the forehead and eyebrows, widening the eyes. Because these muscles connect directly to the skin, every

• DIFFERENT INTERPRETATIONS •

Facial muscles are universal, but the meanings of expressions can be different around the world. A smile signals joy almost everywhere, but in some East Asian cultures it can indicate embarrassment. In addition, a few remote communities interpret a fearful face as a threatening one.

The effect on mood

The facial feedback hypothesis suggests that when the skin moves into a particular expression, such as a smile, the brain receives feedback that reinforces the corresponding emotional state. Studies have shown that forcing a smile can improve mood, while preventing certain expressions – such as frowning – can reduce negative emotions.

This is thought to be one of the reasons why botulinum toxin (Botox) injections (see pages 152–53), which temporarily relax facial muscles, have been linked to improvements in mood. By reducing the ability of the skin to wrinkle into a frown, the brain receives fewer signals associated with negative emotions, which may contribute to a more positive emotional state.

Facial mimicry is a crucial part of social bonding. People often unconsciously mirror the expressions of those around them, an automatic response that helps build connection and rapport. The skin's movement in these interactions plays a key role, allowing for rapid and precise adjustments that make facial communication more fluid and natural. This process is thought to be driven by a specialized network of neurons known as the mirror neuron system, which activates both when we observe an expression and when we produce it ourselves.

Facial expressions do not just communicate emotion – they can also influence how we feel.

WHY DOES MY FACE GO BRIGHT RED WHEN I AM EMBARRASSED?

When you feel embarrassed, your brain has registered a social threat and it flips your "fight-or-flight" switch. Adrenaline is released, your heart beats faster, and blood vessels in certain areas of skin – especially the face, neck, and upper chest – suddenly dilate. Because the facial skin is rich with blood vessels and they sit very close to the surface, that rush of blood shows through as a bright-red flush. This kind of blushing is a very specific, socially tuned response. We do not flush like this when we run up the stairs, even though the heart rate rises. The nerves supplying the tiny muscles in the vessel walls of the face seem particularly sensitive to emotions such as embarrassment, shame, or self-consciousness. Blushing can actually make people seem more sincere and approachable to others, even while it feels mortifying from the inside.

—

CAN MAKING CERTAIN FACIAL EXPRESSIONS CAUSE PERMANENT LINES?

Every time you frown, squint, smile, or raise your eyebrows, underlying muscles contract and pull the skin into folds. In young, well-supported skin, those lines disappear as soon as the face relaxes. Over years, especially in skin that has also been thinned and stiffened by ultraviolet damage and smoking, those same fold lines become etched into the dermis. Collagen and elastin fibres gradually reorganize around the repeated crease, so what started as a movement line becomes a "static" wrinkle that is visible even when your face is completely at rest. That is why we see familiar patterns – vertical lines between the brows in people who frown or concentrate intensely, "crow's feet" lines around the eyes in habitual smilers and squinters, and horizontal forehead lines in people who lift their brows a lot.

—

WHAT CAUSES THE PINS AND NEEDLES SENSATION IN MY SKIN?

When you sit on a limb, you compress the nerve fibres and the tiny blood vessels that feed them. Mechanical compression distorts the nerve, which can briefly block normal message transfer; reduced blood flow then worsens this by disturbing the ion balance (sodium, potassium, calcium) across the nerve membrane. These nerve fibres then start firing spontaneously and out of sync. The brain gets this noisy, disorganized barrage from sensory fibres and experiences it as tingling, prickling, and buzzing – pins and needles.

—

CAN PEOPLE TELL I AM UNWELL FROM MY SKIN BEFORE I FEEL REALLY ILL MYSELF?

Yes, this can often be the case. There are two ways of looking at it – those who are about to be acutely unwell and those who have a condition that takes a while to rear its head. Your skin is closely tied to your circulation, immune system, hormones, and nervous system, so it can change quite quickly when something inside you is starting to go wrong. Friends or family might notice you "don't look yourself" before you feel dramatically unwell: a bit more washed out or pale, a grey or sallow tone, slightly sunken eyes, unusual redness across the cheeks, or a fine, sandpapery rash. There are also slower changes that you might miss because you see your face every day. In fact, when my son had appendicitis, it was the colour of his face that gave me more clues than his abdomen! For slower changes, someone who hasn't seen you for a while may clock that your skin has become drier, that you are scratching more, that your hair is thinner, or that you have lost the underlying plumpness that used to sit in your cheeks. Those gradual shifts can signal chronic stress, sleep deprivation, nutritional problems, endocrine issues, or low-grade inflammatory disease. That said, it is not a precise diagnostic tool and looking "a bit off" does not mean there is necessarily something serious brewing; very significant illnesses can sometimes develop with almost no visible skin change at all.

—

03

Skin through the ages

neonatal and infant skin

In the first year of life, our skin learns how to function. How we care for it in those early months can have long-lasting effects, from the development of a robust skin barrier and microbiome to the prevention of food allergies.

When we're born, our skin is coated in a creamy, white substance called vernix caseosa. This is a mixture of water, sebum (see page 15), and dead skin cells that protects us from the amniotic fluid in the womb and helps us slip through the birth canal.

It continues to work after we're born by supporting our skin barrier and encouraging the right microbes to settle, but most importantly it reduces transepidermal water loss (TEWL). This is the amount of water that naturally escapes from the skin into the air. As a result, vernix should not be washed off straight away. The World Health Organization advises waiting at least 6 hours before the first bath, giving it time to absorb.

Caring for young skin

When it comes to caring for infant skin, bathing should be short - only a few minutes in warm (not hot) water, ideally two to three times a week. In between, wiping with a soft flannel or washcloth is more than enough.

Traditional soap is too alkaline for delicate skin and disrupts the skin's pH and lipid layers. Instead, choose a soap-free cleanser (known as a syndet) that is unfragranced and pH-balanced at around 5.5. It's tempting to choose beautifully packaged, scented, bubbly baby products but young skin doesn't need those. In fact, fragrance, bubbles, and botanical additives are common causes of skin irritation in the first year of a baby's life. The simpler the cleanser, the better.

THE PH SCALE AND BABY SKIN

The pH scale measures acidity and alkalinity. At birth, the skin surface pH is close to neutral, but over the next few months it gradually acidifies. The maturation of this layer (known as the acid mantle) during the first year of life is an important step in the development of a healthy skin barrier.

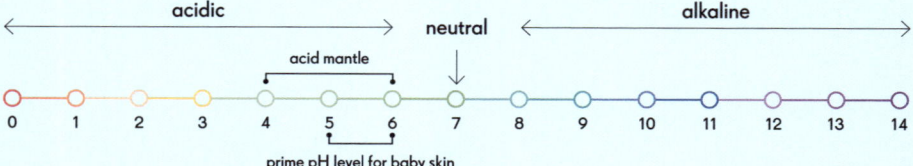

THE MOST COMMON SKIN CONDITIONS IN OUR FIRST YEAR

The majority of baby skin concerns are rashes or issues with skin that is too dry or oily. Most are temporary, manageable, and improve with gentle care and protection.

CONDITION	SYMPTOMS	CAUSE	WHAT TO DO
MILIA	tiny white bumps, usually on the nose or cheeks	trapped keratin in immature pores	nothing – these resolve on their own, though a gentle wipe with a damp cloth can help
HEAT RASH (MILIARIA RUBRA)	small red bumps, often in skin folds, on the face, neck, or back	blocked sweat glands during hot weather or after viral illness	cool the skin with a damp cloth, dress lightly, avoid overheating. It clears quickly
NAPPY RASH (DIAPER DERMATITIS)	red, irritated skin in the nappy area	moisture, friction, and prolonged contact with urine or faeces	change nappies frequently, use fragrance-free wipes or just water, apply a barrier cream, and allow nappy-free time
CRADLE CAP (INFANTILE SEBORRHOEIC DERMATITIS)	waxy yellow scales and redness on the scalp, sometimes spreading to the eyebrows or ears	likely related to maternal hormones stimulating sebum production and the yeast *Malassezia* thriving on the skin's oils	usually self-resolving. Gentle shampoo and brushing may help. Apply mineral oil or white petrolatum overnight to loosen scales. If persistent, a mild antifungal or steroid may be used under guidance – but in most cases it just needs time
ATOPIC ECZEMA	dry, red, and itchy patches on the cheeks, and later the creases of elbows, knees, or neck	a combination of genetic predisposition and environmental triggers, linked to an impaired skin barrier and immune imbalance	daily moisturizing and early management are key as doing so promptly will treat more than just the skin. Emerging research suggests that untreated eczema may increase the risk of developing food allergies by allowing allergens to enter the body through the skin. Encouragingly, with gentle care, many babies grow out of it

teenage skin

As teenagers, our skin is impressionable in every sense of the word.
It's biologically responsive to hormones, environmentally reactive, and
socially shaped by a digital world with very curated beauty ideals.

Today, teenagers are navigating skincare not just in pharmacies or bathrooms, but through influencers, trends, and algorithms. As we explore skincare in our teenage years, our skin needs more than just access to products – it needs protection from pressure, misinformation, and unrealistic expectations.

During puberty, our skin undergoes a series of hormonally driven changes, one of the most visible being the change that occurs in the hair in our pubic region and armpits. Under the influence of increasing circulating androgens (sex hormones), particularly dihydrotestosterone (DHT), the fine, light vellus hairs that cover most of our body transform into terminal hairs. These hairs are thicker, coarser, darker, and their follicles are more deeply rooted in the dermis. This transition is a key marker of sexual maturation.

Pubic and armpit hair

Pubic hair typically first appears along the mons pubis and progresses in a predictable pattern, known in clinical terms as the Tanner staging system. Early pubic hair is often lightly pigmented and fine (Tanner stage 2), becoming progressively curlier, denser, and more pigmented as puberty advances (Tanner stages 3–5). For some, pubic hair may eventually extend up to the linea alba on the abdomen (forming a "trail") and the upper legs and buttocks.

The teenage years may be our first real experience of feeling out of control in our own skin.

-)-)-)-
.·.··.·.

The development of hair in the armpits follows a similar androgen-mediated transformation and typically occurs slightly after the appearance of pubic hair. The follicles in these areas also become active apocrine sweat glands. Located in the armpits and groin, they secrete a thick, protein-rich fluid that bacteria break down into body scent. They contribute not only to hair growth, but to the onset of "teenage" body odour.

Excess sebum

One of the more problematic changes that can occur happens in our sebaceous glands, which, fairly quiet until now, suddenly come to life. They grow larger and begin pumping out sebum, the skin's natural oil. This oil helps protect and moisturize our skin, but when produced in excess, it mixes with dead skin cells and clogs our pores. That's when we begin to notice spots and blackheads, and for some of us, more inflamed acne. Our skin starts to feel oilier, particularly across the forehead, nose, and chin. Acne is the most common skin condition we experience during our teenage years, and we'll explore it in greater detail on pages 68–69.

Dandruff and acne

At the same time, the microbiome of our skin also shifts and the increase in oil provides a richer environment for certain bacteria, particularly *Cutibacterium* acnes, which play a role in acne development.

Other microbes such as *Malassezia*, a type of yeast, also thrive in the oilier setting, particularly on our scalp and the oilier parts of our face, and may trigger seborrhoeic dermatitis, also known as dandruff. This is a by-product of the inflammation created between the oil and the yeast that is feeding on it and is extremely common in teenage skin. Tips on its management can be found on page 162.

in your twenties

You will never have healthier skin than when you are
in your twenties. Yet while the skin's surface may look
youthful, subtle biological changes are already underway.

By the time most of us reach our twenties, our skin is typically functioning at its peak. The epidermis maintains a healthy turnover rate, the dermis is rich in collagen and elastin, and the skin barrier is generally resilient. This is often the time when skin appears firm and well-hydrated, supported by abundant hyaluronic acid and efficient repair mechanisms.

Signs of change

Nevertheless, from around the age of 25, we begin to see the earliest measurable signs of intrinsic ageing (see page 52). In clinical terms, we may start to observe early fine lines around the eye area, slight dullness due to slower cell turnover, and skin wrinkling, especially in the context of fatigue, stress, or environmental exposure.

This is also the time of life when extrinsic ageing (see page 52) begins to assert itself more visibly. In patients who have had significant UV exposure – particularly from incidental daily sun or the use of sunbeds in adolescence – early photodamage may present as uneven pigmentation, subtle irregular brown freckling, or roughened skin texture. Prevention at this stage is far more effective than reversal later on. Consistent daily application of a sunscreen protecting against UVA and UVB remains the single most important intervention I can recommend.

• LAYING FOUNDATIONS •

If a patient in their twenties sits in front of me, I remind them of a key point: their skin
is strong, and repairing itself effortlessly. The danger for many is doing too much.
Don't overload your skin. This decade is about laying secure foundations.

-)-)-)-

Hormonal activity

Hormonal shifts can still play a role in our twenties, particularly for women. Acne spots may persist or reappear, often clustered around the jawline, chin, and neck. Conditions like PCOS (polycystic ovary syndrome) may declare themselves more clearly now, with a combination of irregular periods, acne, oilier skin, or excess facial hair. We'll explore this in more depth in a later chapter (see pages 184–85), but it's worth noting how often it intersects with dermatological concerns in this life stage.

Starting a skincare regime

What also shifts in this decade is behaviour, with more advanced active products becoming part of a skincare regime. While some introduce these appropriately, others present with signs of barrier dysfunction from overuse of exfoliants, harsh cleansers, or the layering of incompatible products (see pages 118–19).

While the skin at this age is often at its physiological peak, repeated inflammation from chronic skin conditions such as eczema or acne can begin to leave its imprint in the form of post-inflammatory pigmentation, scarring, or areas of thickened skin where the barrier has been broken down over time.

Mostly, our skin in our twenties is adaptable – it still has the power to regenerate well, respond quickly to good routines, and recover from setbacks. The choices we make now about sun protection, stress management, nutrition, and skincare may not show immediate rewards, but they will compound over time.

Many twenty-somethings begin using active skincare ingredients for the first time – often with little guidance.

-)-)-)-

how our skin ages

Our skin ages in two ways, both happening at the same time. Intrinsic ageing is already predetermined and is written in our DNA code, while extrinsic ageing is related to our lifestyle and the choices we make.

Intrinsic ageing

Intrinsic ageing is a natural, gradual decline in skin structure and function. With time, collagen and elastin production decreases as fibroblasts, the cells responsible for making them, slow down. This means the scaffolding of our skin is much weaker. The rate at which the skin cells turn over in the epidermis drops, leaving our skin duller in appearance but also more fragile, with slower wound-healing and a diminished immune response. For women, postmenopausal hormone imbalances amplify these changes, and oestrogen deficiency contributes to the thinning of the epidermis and dermis.

Extrinsic ageing

Extrinsic ageing, in contrast, is largely preventable. It's caused by external assaults on the skin, including ultraviolet (UV) light, pollution, poor nutrition, and cigarette smoking. Ultraviolet A (UVA) light, which makes up around 95 per cent of the light from the sun's rays, is far and away the most damaging of these. It initiates a cascade of cellular events that generates free radicals (see page 20) – these molecules damage DNA, further degrading collagen and elastin, and impairing immune function. The result is photo-ageing: a leathery texture, deep wrinkles, pigment irregularities, broken capillaries, and skin cancers. Ultraviolet B (UVB) light rays from the sun also cause damage, through burning, which leads to mutations within the cells that potentially lead to a cancerous change.

Smoking compounds these effects by impairing blood flow and nutrient delivery, degrading collagen, and accelerating wrinkle formation. Studies show a close relationship between smoking cigarettes and facial wrinkling. Skin in smokers often appears dull, sallow, and prematurely aged, and most of us dermatologists can tell if you are a smoker simply by looking at your skin.

Inflammaging and "zombie" cells

One of the most intriguing and impactful concepts in skin ageing is inflammaging, which is a chronic, low-grade inflammatory state that develops as we get older. A key contributor to this process is the accumulation of senescent

-)-)-)-

(ageing) cells, often called "zombie" cells. These are old, damaged cells that no longer divide or function properly, but instead linger in the skin, releasing inflammatory signals and enzymes that harm neighbouring cells and promote tissue breakdown.

Over time, zombie cells act like bad apples in a fruit bowl, spreading dysfunction and accelerating the visible and structural signs of ageing. While our skin constantly renews itself, the growing presence of these senescent cells can tip the balance towards degeneration rather than repair.

Understanding and eventually targeting inflammaging and zombie cells holds the key to healthier ageing skin; senolytic drugs – applied to the skin (topically) or taken internally (systemically) – are an active area of research. An already widespread drug that has shown promise at targeting senescence is metformin, a drug commonly used in diabetes. Watch this space!

THE DEVELOPMENT OF SENESCENT CELLS

Senescence begins when a damaged cell stops dividing. The proteins it releases attract immune cells that should eliminate the senescent cell; however, in ageing skin, senescent cells can accumulate rapidly as immune clearance fails to keep up.

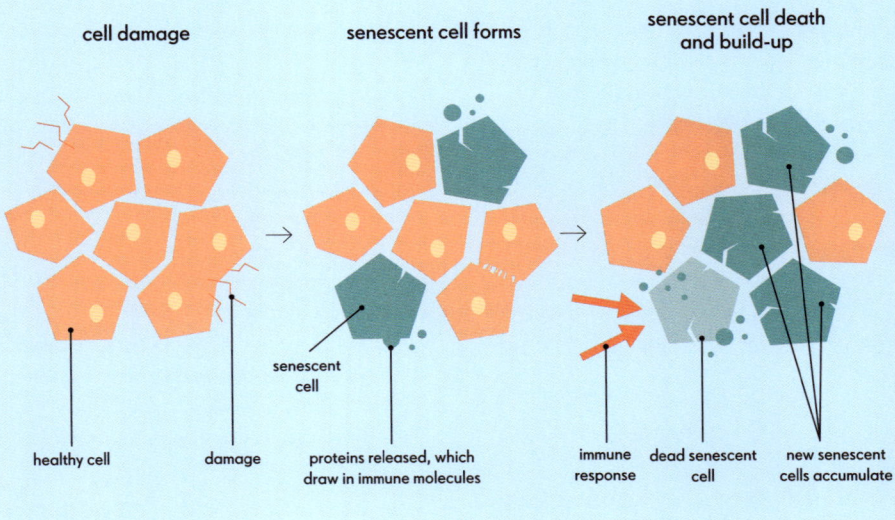

cell damage → senescent cell forms → senescent cell death and build-up

senescent cell

healthy cell · damage · proteins released, which draw in immune molecules · immune response · dead senescent cell · new senescent cells accumulate

pregnancy

Pregnancy is a time of extraordinary transformation.
Changes to the skin can sometimes be concerning,
but most are entirely physiological and reversible.

Pigmentation

If you're pregnant and noticing new patches of pigment, you're not imagining it. Nearly 90 per cent of women experience some increase in pigmentation, especially those with darker skin tones. Nipples, armpits, inner thighs, and genitals may darken. A line may appear down your belly – *linea nigra* – where once there was the faint *linea alba*. Moles, scars, and freckles often deepen in colour too, and roughened spots may appear, especially on the torso.

Then there's melasma (see pages 76–77), sometimes called the "mask of pregnancy". It often shows up as symmetrical brownish patches on the cheeks, forehead, or upper lip, and can feel unfairly persistent. Caused by hormonal shifts and sun exposure, it tends to return unless SPF (see pages 126–27) is used correctly and consistently. Proper sun protection really is the only long-term strategy that works.

Stretch marks

Stretch marks – *striae distensae* – often appear in the second or third trimester. They tend to show up as soft, reddish lines across the abdomen, breasts, thighs or buttocks, revealing just how much your skin is adapting. These marks result from both stretching and genetics – some of us are simply more prone than others. Over time, they fade. What begins as red or purple usually settles into paler, slightly indented streaks. No cream or oil can reliably prevent or erase them, but keeping skin well moisturized may ease discomfort and support elasticity.

Hair

One of the perks of pregnancy is your hair – it can feel thicker, shinier, and more abundant. This is due to rising oestrogen levels, which hold hair in its growth phase for longer. Yet after delivery, as hormones shift, things change again. What often follows is *telogen effluvium* – a sudden, noticeable shedding that can start one to five months postpartum. It's unsettling, especially when clumps appear in the shower or brush, but it's temporary. The follicles aren't damaged, and regrowth usually begins on its own. Typically, the hair will return to its usual pattern within a year.

-)-)-)-

Vascular changes

As blood volume and vascular sensitivity increase, you may notice more visible blood vessels than before. One common example is spider angiomas – small, bright-red spots with tiny vessels fanning out from the centre like legs. These often show up on the face, chest, or neck, and affect up to two-thirds of pregnant women.

Skin diseases in pregnancy

Polymorphic eruption of pregnancy (PUPPP) is the most common pregnancy-specific skin condition. It usually affects first pregnancies in the third trimester, starting as intensely itchy red bumps in abdominal stretch marks. The rash may spread to the thighs, buttocks, or arms. It looks dramatic but poses no risk to the baby.

Prurigo of pregnancy appears earlier, often in the second trimester, with scattered, itchy bumps on the arms and legs. It may last throughout pregnancy and recur in future ones, particularly in those with atopic skin. Emollients and topical steroids are usually sufficient.

Pruritic folliculitis of pregnancy resembles acne or folliculitis but without infection. It affects the trunk, is itchy, and usually clears postpartum. Mild topical treatment usually works well.

Pregnancy can also alter chronic skin conditions. Autoimmune diseases such as lupus or psoriasis may flare or improve, while atopic eczema often worsens.

ANATOMY OF A STRETCH MARK

Stretch marks appear when the skin stretches too fast for the fibres within it. They start off as red marks and then over time fade to pale, silvery lines.

epidermis • collagen & elastin • thinner area of dermis • epidermis • collagen & elastin • dermis • blood vessel • hypodermis • blood vessel • hypodermis • dermis

healthy skin • stretch mark skin

menopause

During menopause, the reduction in the levels of the hormones oestrogen and progesterone contributes not only to symptoms such as hot flushes, but also to notable skin changes.

Menopause represents a single day in a woman's life when monthly periods have ceased for one year. It typically occurs between the ages of 45 and 55, but for up to 10 years before menopause there are hormonal fluctuations as oestrogen and progesterone levels fall.

Oestrogen reduction

Oestrogen has its hand in almost every part of the skin. It helps maintain the lipid barrier, supports hydration, and keeps collagen and elastin production ticking over. So, when levels begin to fall, the skin is affected, and one of the most noticeable changes is the loss of collagen. In fact, about a third of our dermal collagen disappears within the first five years after menopause, and the decline continues slowly after that. This isn't just about wrinkles – collagen is what gives skin its structure and bounce, and without it, we see thinning, sagging, and skin that is much more fragile.

Our epidermis also relies on oestrogen and, as levels drop, the skin barrier becomes less efficient. Water escapes more easily, leading to dryness (xerosis) and itching (pruritus), two of the most common skin complaints during menopause. Eczema or dermatitis may flare too, especially if you've been prone before. While hormone replacement therapy (HRT) may improve skin hydration for some, daily care makes a real difference. Low-pH emollients, short lukewarm showers, fragrance-free cleansers, and even a bedroom humidifier in winter can help restore balance and comfort.

Acne and hot flushes

For many women, acne returns during the perimenopause, due to a relative rise in androgens – the so-called male hormones – as oestrogen levels fall. It tends to break out around the lower face and jawline. This type of acne calls for a different approach than in younger skin. Topical retinoids and azelaic acid are effective, but they need to be used in more hydrating formulations. Oral options such as spironolactone or low-dose isotretinoin tend to work better than antibiotics and are often well tolerated in this age group.

Hot flushes are also common, affecting about 75 per cent of menopausal women. These sudden waves of heat, caused by widening blood vessels in response to decreasing oestrogen, can be distressing. HRT is the most effective treatment, but non-hormonal options can also make a real difference.

-)-)-)-

Hair issues

Many women notice hair thinning through the top of the scalp, even though the hairline itself stays put. This pattern of hair loss often speeds up as oestrogen levels fall. While hormones like androgens can play a role, it's the loss of oestrogen's support that seems to matter most.

There are things that help. Topical minoxidil (available as a 2% solution or 5% foam) can slow the process and encourage regrowth, though patience is key. For some women, especially if the hair loss is distressing or progressing quickly, oral treatments such as spironolactone or finasteride may be considered. Nutritional supplements can support general hair health, but they're rarely game-changing on their own.

A different kind of hair loss is frontal fibrosing alopecia, which is usually seen around or after menopause. This condition causes the hairline to slowly recede and the eyebrows to thin or disappear. It's a scarring type of hair loss, which means it can be permanent if not caught early. If you're noticing changes such as this, it's worth seeing a dermatologist as soon as you can.

Vulval skin

The vulval skin can become thinner, drier, and more sensitive after menopause, and for some women, this can lead to itching, discomfort, or pain during sex. In this instance, a small amount of topical oestrogen can make a big difference.

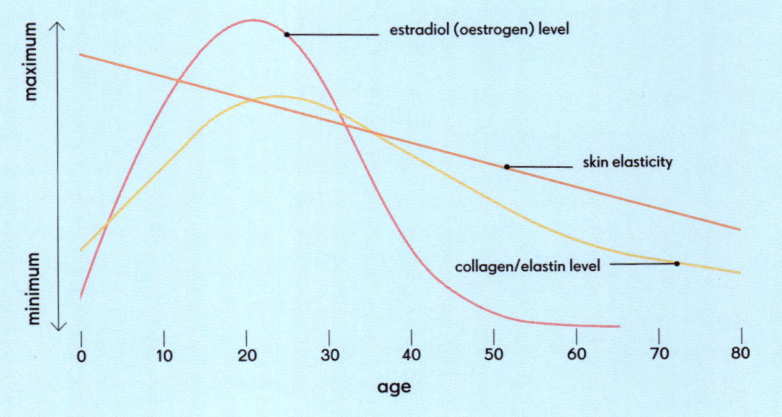

OESTROGEN AND COLLAGEN REDUCTION DURING MENOPAUSE

Estradiol is the most potent type of oestrogen hormone. Its decline to low levels during the menopause, from around 45 years old, affects the amount of collagen and elastin in the skin.

-).-).-)-

mature skin

From thinning to dryness, bruising, and a susceptibility to infection, our skin undergoes radical changes as we get older. As a result, mature skin deserves thoughtful, proactive care.

The most consistent change anatomically in mature skin is thinning across all the layers. The outer layer becomes more fragile, while the collagen-rich middle layer loses volume. With less support, skin becomes looser and fine lines deepen into folds. You might notice a crêpe-paper texture on the neck, arms, or eyelids. Wounds take longer to heal, and bruising or pressure injuries become more likely.

Dry skin

Dryness is one of the most common complaints in mature skin. As we age, oil production slows, and natural moisturizing factors, like ceramides and amino acids, decline. This leaves skin feeling tight, flaky, or itchy, especially in colder months. A weakened skin barrier is part of the story too, because it not only lets moisture escape more easily, but also becomes slower to repair after irritation or injury. As a result, the skin is more reactive to things such as central heating, rough fabrics, or even products you used to tolerate.

Regular daily habits make the difference here. Use gentle, soap-free cleansers and moisturize your body after washing.

Bruising and tears

A significant structural change occurs in mature skin – the flattening of rete ridges. These dividing lines between the epidermis and the dermis are wavy to enable more nutrients to pass through, but also hold the epidermis in place. Flat rete ridges make the skin more fragile and it loses some of its mechanical strength – the skin breaks more easily and takes longer to heal. Over time, small blood vessels are also more fragile, which leads to easy bruising.

Conditions affecting mature skin

There are some skin conditions that become more prevalent – or more troublesome – with age.

- **Herpes zoster (shingles):** If you had chickenpox as a child, the virus lies dormant. Later in life – typically after 50 – it can reactivate as shingles: a painful rash made up of small, blistering patches. The pain can linger long after the rash fades, known as post-herpetic neuralgia. The good news is that there's a vaccine that offers real protection, and starting treatment early can make the episode shorter and less severe.

-)-)-)-

- **Pruritus (itch):** This is often linked to dryness, but can also occasionally be nerve-related and reflect an internal issue such as kidney disease. Scratching can lead to secondary eczema or lichen simplex chronicus – thickened, scaly plaques from habitual rubbing.

- **Eczema and dermatitis:** These conditions often flare up in later life due to a weakened skin barrier and immune shifts. What you tolerated for decades may now trigger an itchy rash.

- **Actinic keratoses and skin cancer:** Years of sun exposure can leave their mark as rough, scaly patches on the skin called actinic keratoses. These are tell-tale signs that the skin is at increased risk of developing skin cancers.

- **Seborrhoeic keratoses:** These harmless, warty-looking growths pop up with age. Though annoying, they are nothing to worry about.

- **Fungal nail infections:** These crop up more often because circulation slows and nails grow more slowly, giving fungi more time to settle in.

Taking care of older skin

What used to be dismissed as vanity is now viewed as smart skin maintenance. Retinoids, injectables, or energy-based devices (see pages 155–56) can help preserve structure, support the skin barrier, and boost its ability to repair. It's not about chasing youth, but about keeping skin healthy and resilient as the years go by.

THE SKIN AS IT AGES

Wrinkles, pigmentation changes, and loss of resilience in older skin are due to the thinner epidermis, disorganized collagen, and reduction in blood flow supply to the skin.

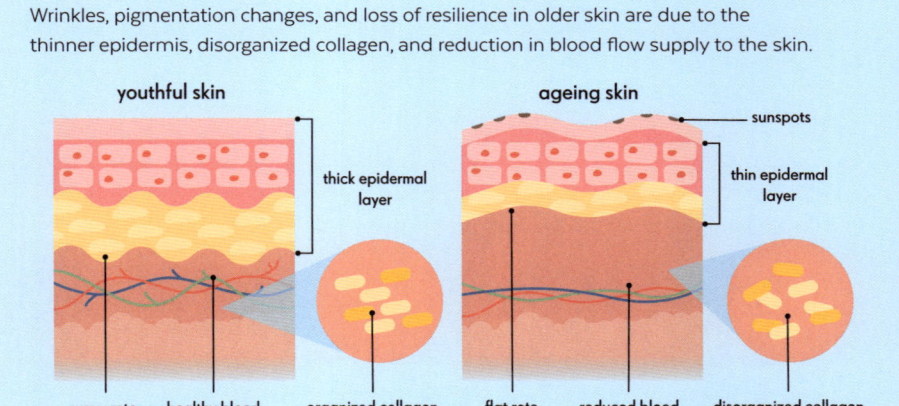

youthful skin

thick epidermal layer

wavy rete ridges

healthy blood vessel formation

organized collagen and elastin

ageing skin

sunspots

thin epidermal layer

flat rete ridges

reduced blood vessel formation

disorganized collagen and elastin

special areas of focus

Not all areas of the skin are created equal. Some regions of the body are built for heavy lifting, whereas others are made for exquisite sensitivity.

The skin of the palms, soles, nails, and mucosal sites – like the eyes, mouth and in the genital area – is structurally and functionally distinct, and each behaves in its own way.

Palms and soles

Palmar (palm of the hand) and plantar (sole of the foot) skin is thick, robust, and free of hair follicles. The stratum corneum – the outermost layer – is much denser here than elsewhere, designed to resist mechanical friction. With age, even this hardy skin thins and loses cushioning, especially on the soles. Fat pads shrink, making walking less comfortable and predisposing to calluses, corns, and fissures.

Sensory perception also declines, increasing the risk of unnoticed injury.

Care tips: Hydration is key. Thick emollients containing urea or lactic acid can help soften and restore cracked heels or dry palms. Regular podiatric care becomes more important with age, particularly for those with diabetes or reduced mobility. And never underestimate the benefits of footwear that fits well.

Nails

Nails are modified skin appendages made of hard keratin. They grow slowly throughout life and, rather like tree rings, they reveal clues about our systemic health. With age, nails can become

---

• READING THE NAILS •

Nails can often give clues to changes in our general health. For example, anaemia can cause spoon-shaped nails and psoriasis can lead to splitting and pits appearing. Transverse grooves across the nail plate represent a temporary halt in nail growth at the time of a fever or illness. You can even measure this distance to see how long ago the illness occurred!

more brittle, ridged, and prone to fungal infections. Blood flow to the nail matrix reduces, slowing growth and healing.

Care tips: Protect nails from trauma with gloves during manual tasks. Keep them trimmed and filed. If nails become thickened, discoloured, or lift from the nail bed, seek a medical review – it's not always "just age".

Genital and mucosal sites

The vulva, penis, scrotum, perianal region, and mucosal surfaces (such as the lips and inside the mouth) have unique structural characteristics: thinner skin, more vascularity, and often greater sensitivity to hormonal changes.

In women, the menopausal decline in oestrogen leads to thinning of the vulval and vaginal epithelium, reduced lubrication, and increased fragility – this is known as genitourinary syndrome of menopause (GSM). Itching, burning, fissuring, and discomfort during intercourse are common, but not inevitable.

In men, thinning of the genital skin and mucosa can also occur, and chronic inflammatory conditions such as lichen sclerosus may emerge. Mucosal immune surveillance – where the immune system constantly monitors mucosal surfaces – also diminishes with age, increasing the risk of persistent infections or delayed healing.

Care tips: Choose fragrance-free, pH-appropriate cleansers for these areas. Avoid tight-fitting clothing and consider barrier creams if prone to irritation. For postmenopausal vulval dryness or irritation, topical oestrogen may help.

Lips and oral mucosa

Lips are technically part of the mucocutaneous junction where the skin transitions into mucosa. They lack sebaceous glands, making them vulnerable to dryness and cracking. With age, loss of volume, fine lines, and thinning skin become more prominent, and dental health can affect the surrounding tissue.

Care tips: Regular lip balm with SPF is more than a comfort – it's protection. For persistent soreness, cracks at the corners (angular cheilitis), or white patches, don't delay getting your lips checked.

-)-)-)-

loving the skin you're in

Your skin tells a story. It is yours to understand, shape, and accept. Ultimately, it's not the appearance of your skin that defines your experience – it's how you feel about it.

Every person carries a unique skin story. For some, it's a tale written in freckles or birthmarks that have always been there. For others, it's the echo of inflammation, acne scarring, eczema flare patterns, or pigment changes after trauma. Some skin changes are deliberately chosen, like tattoos or piercings. Thinking back to how our skin ages, it is important to remember that at any of these stages, healthy skin is the goal, so whatever stage you're at in your skin journey, it's worth remembering: this is the skin that's carried you through every season of life so far. It deserves your care and respect.

The power of perception

Studies in psychology have consistently shown that emotional impact is not dictated by the objective severity of a skin condition. Instead, it's our perception of it that matters. One person may feel empowered by a visible difference, while another might struggle deeply with something less noticeable to others. There's no right or wrong here, just individual reality. Our thoughts, feelings, and actions create a loop. The way you think about your skin shapes how you feel about it. That emotional response then influences what you do next. But this loop can be interrupted and reshaped. By changing your actions – such as the way you care for your skin or speak to yourself – you can begin to shift how you feel. Even small steps count.

Caring for our skin means balancing change and acceptance.

How to change your approach

Start with structure. Develop a routine that supports your skin health and gives you a sense of agency. Keep it simple – a gentle cleanser, a hydrating moisturizer, a broad-spectrum sunscreen. Write it down if that helps. Keep the products visible and easy to access. Repetition brings comfort and builds confidence.

Watch your language. When you look in the mirror, notice how you speak to yourself. If your inner voice is harsh or critical, pause and redirect. Say something kind and reassure yourself. If the same thoughts keep returning, try a technique known as "thought stopping" – acknowledge the unhelpful thought, then gently let it go. Shift your attention to something you enjoy such as music, a favourite book, or even a friend's voice.

Some people find it helpful to wear a soft hair band or bracelet. When negative thoughts about skin come to the surface, a gentle touch of that item can act as a cue. It's a physical reminder to change the channel, to turn down the volume of self-criticism, and choose a different response.

Rethinking social media

Social media can be a powerful connector, but it also creates a distorted view of reality. Most people only share filtered photos. Unfiltered skin, uneven tone, or natural texture are rarely seen. Limiting time online, curating who you follow, and taking breaks when needed can help reset your perspective.

Your skin is not always flawless, but it is always functional. It does more for you than any filter ever could.

HOW CAN I PREVENT AND TREAT NAPPY RASH WITHOUT MAKING MY BABY UNCOMFORTABLE?

Most nappy rash is caused by a mixture of prolonged contact with wee and poo, friction from the nappy, and a warm, humid environment. So, in order to prevent it, change nappies promptly when they are wet or soiled, use lukewarm water or very mild, fragrance-free wipes, and pat rather than rub the skin dry. A thin, disposable nappy that wicks moisture away from the skin is usually more comfortable than thick, non-absorbent layers. A simple barrier cream is your best friend – after each nappy change, apply a generous layer of a fragrance-free ointment (for example, zinc oxide or a petrolatum-based barrier) so that the next wee or poo hits the cream, not the skin. You should still be able to see a bit of white cream at the next change; if it has completely disappeared, use a thicker layer. Nappy-free time on a towel or waterproof mat once or twice a day gets air to the area and makes babies very content. Parents often notice that nappy rash seems worse when a baby is teething. The teeth coming through do not damage the skin in the nappy area, but this stage of life is full of mild viral infections and looser stools, and those are exactly the things that make nappy rash flare.

———

HOW MUCH SKIN-TO-SKIN CONTACT SHOULD MY BABY HAVE? AND DOES IT REALLY HELP THEIR SKIN OR IS IT MAINLY FOR BONDING?

For healthy term babies the evidence supports immediate skin-to-skin contact after birth for at least about an hour, and ideally until the end of the first breastfeed. After that first hour, there is no upper limit. Ongoing skin-to-skin in the first days and weeks (for feeds, settling, naps on chest) is encouraged "for as long as parents wish"; more contact is associated with better breastfeeding rates and calmer babies. Bonding is a big part of it, but high-level evidence shows very concrete physiological and behavioural benefits of skin-to-skin contact – babies are more likely to breastfeed, breastfeed longer, and have more stable temperature, blood glucose, heart rate, and breathing in the hours after birth. They also cry less and display fewer signs of stress.

———

HOW OFTEN DO TEENAGERS NEED TO WASH THEIR FACE?

Most teenagers are fine washing their face twice a day – once in the morning and once in the evening – plus an extra gentle rinse after sport or very sweaty activity. The aim is to remove excess oil, sweat, pollution, and sunscreen without stripping the skin. Use a mild, non-fragranced cleanser, not bar soap or harsh scrubs. In the morning, a quick cleanse gets rid of overnight oil and products; in the evening, take a little more time to remove make-up and sunscreen, and then cleanse. Washing more than twice a day, or scrubbing hard, will not "clean out" acne – it just irritates the skin barrier and can actually make breakouts, redness, and oiliness worse. You also can't scrub away blackheads as they are deep in the pore. What ends up happening is that the pore becomes more inflamed, and instead of being a blackhead it may move to being a larger redder acne spot.

—

WHAT CAUSES MY SKIN TO AGE THE QUICKEST?

The things that age your skin fastest are those that repeatedly chip away at its structure and repair systems, day after day, for years. Top of the list, by a very long way, is UV (ultraviolet) light. Unprotected time in strong daylight – not just on holiday, but the accumulated hours of everyday life – breaks down collagen and elastin (tangling them into a yellow, stiff "solar elastosis"), mutates DNA, and scrambles pigment. That is why you often see a stark contrast between sun-exposed and covered areas. Smoking and pollution add their own, very recognizable signature: constant oxidative stress, narrowed blood vessels, starved fibroblasts, a sort of grey, flattened look to the skin, and deeply etched lines around the mouth and eyes. On top of that, there are less obvious lifestyle factors that work more slowly but just as surely to age your skin: chronic poor sleep, long-term unmanaged stress, a diet that is relentlessly high in ultra-processed foods and low in the building blocks your skin needs, and uncontrolled medical conditions such as diabetes. Genetics and hormones absolutely set the background, but in clinic I see time and time again that the biggest accelerators are preventable ones.

—

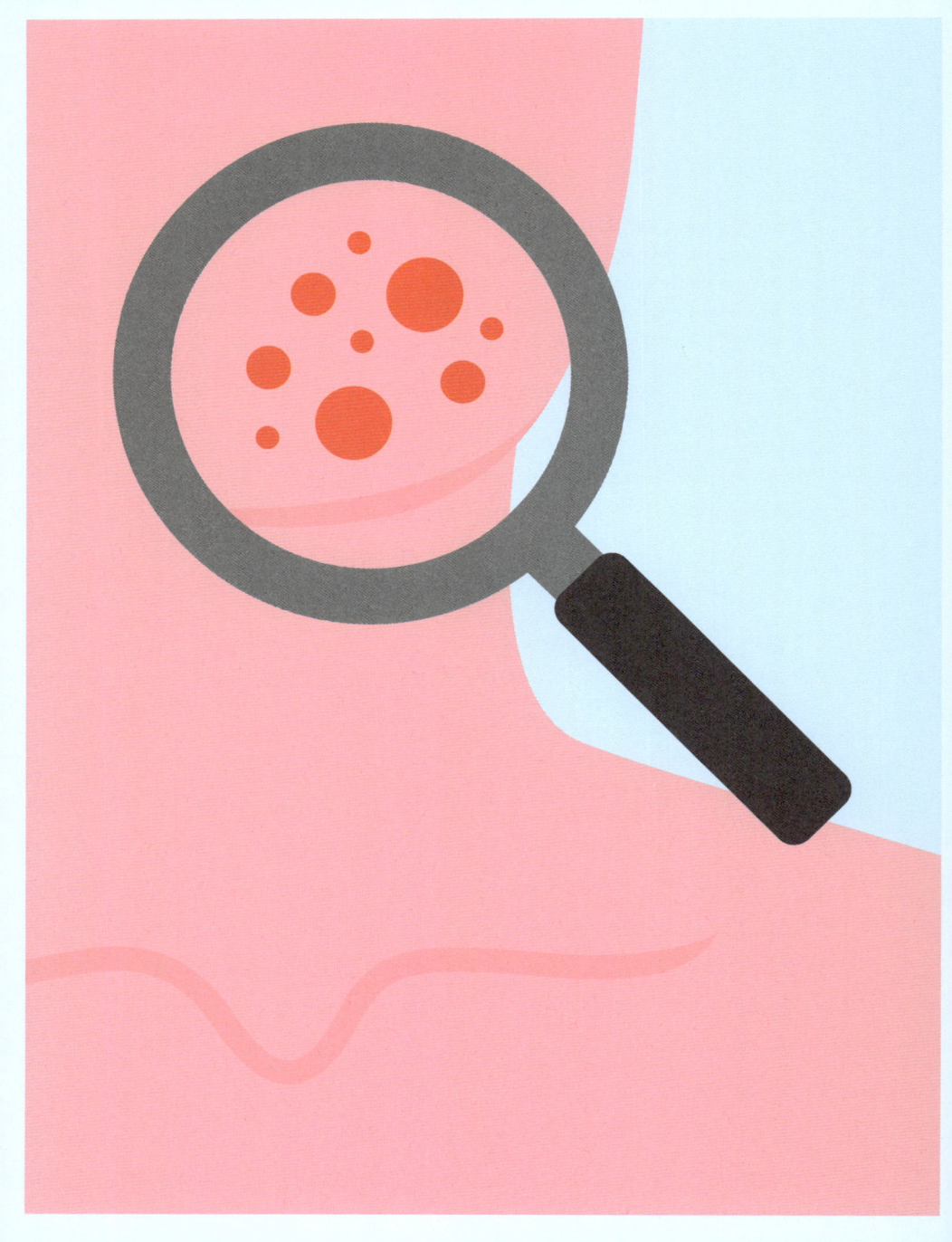

04

Conditions uncovered

acne

Acne is one of the most common inflammatory skin conditions in the world. It affects people of all ethnicities and is most prevalent during adolescence, typically between the ages of 15 and 20.

What makes acne particularly tough is that it's so visible. Whether it's blackheads, red bumps, pustules, or scarring, these changes can take a toll on confidence – the emotional impact can be just as significant as the physical symptoms.

How acne develops

Acne starts deep in the tiny structure where each hair grows – the pilosebaceous unit, made up of the hair follicle and its oil gland. Four main processes are at play: too much oil (sebum), a build-up of sticky skin cells that block the pore, an overgrowth of *Cutibacterium acnes* bacteria, and inflammation.

It often kicks off during puberty, when rising androgens (hormones produced by the adrenal glands) tell the oil glands to ramp up. If the pore gets blocked with the sticky skin cells, the trapped oil becomes the perfect environment for bacteria to multiply, triggering inflammation – that's when you start to see redness, swelling, and tenderness.

You might have a few blocked pores, or widespread sore spots across your face, chest, or back. In darker skin, acne often leaves pigmentation marks. Even mild acne can leave behind scarring in some people, especially if inflammation lingers.

Certain things can make acne worse: stress, medications such as corticosteroids, hormone shifts before a period, changing of an oral contraceptive pill, or heavy or pore-clogging cosmetics. Heat, friction (think helmet straps or mask-wearing), and humidity can also play a role. Although it's tempting, picking or squeezing spots usually makes inflammation worse and raises the risk of scarring.

Treating acne

With so much skincare advice out there – social media, blogs, friends – it can be hard to know what actually works. You might feel like you've tried everything, but sticking with evidence-based treatments is key. For most people, the starting point is a topical medical retinoid. These vitamin A-based creams or gels (like tretinoin or adapalene) help prevent tiny plugs – called comedones – from forming in the pores. Think of them as pore-unblockers that also help your skin shed more evenly. They reduce inflammation and keep things clear; used consistently, they can make a real difference.

If acne is more inflamed – red, sore, or pus-filled – oral antibiotics might be prescribed in the short-term to calm things down. Doxycycline and

-)-)-)-

lymecycline are common options, but they work best with a retinoid – on their own, the benefits often don't last.

For females with monthly flare-ups, or adult acne, hormones may play a part. Certain contraceptive pills or a medicine called spironolactone can help by reducing the oil-stimulating effect of androgens. These treatments often work well in the long term. Changes don't happen overnight, and it can take around 12 weeks before a real change can be noticed clinically, so don't give up or swap treatment too quickly.

Solutions for acute acne

When acne is severe, especially if it causes deep spots or scarring, it might need a different kind of help. Oral isotretinoin is a strong retinoid taken as a capsule. It works by shrinking oil glands, calming inflammation, and stopping new blockages from forming. For many, it clears acne long-term – even permanently – and when used appropriately, it's often life-changing.

Skincare regime and diet

There's growing interest in how diet affects acne, though the evidence is still mixed. High-glycaemic foods and dairy – especially skimmed milk – may trigger breakouts in some people. It's worth paying attention to your skin's response, but strict restrictions usually aren't needed. Instead, focus on a balanced diet (see pages 98–99), consistent gentle skincare (see pages 118–19), and the medical treatments prescribed for you.

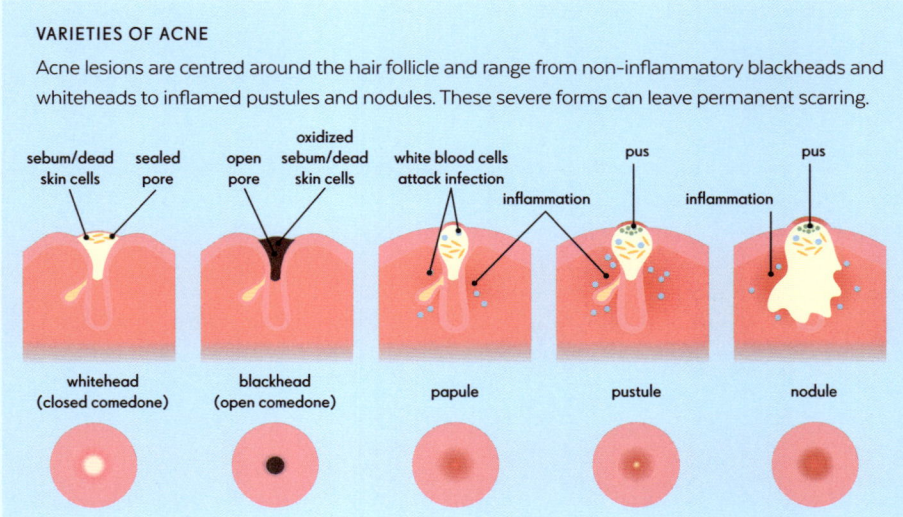

VARIETIES OF ACNE

Acne lesions are centred around the hair follicle and range from non-inflammatory blackheads and whiteheads to inflamed pustules and nodules. These severe forms can leave permanent scarring.

sebum/dead skin cells   sealed pore   open pore   oxidized sebum/dead skin cells   white blood cells attack infection   inflammation   pus   inflammation   pus

whitehead (closed comedone)   blackhead (open comedone)   papule   pustule   nodule

-)-)-)-

eczema

Eczema is a common skin condition, but it's not trivial. It demands empathy and proven evidence-based care to reduce it – not just the severity of the disease, but the secondary derivatives of it.

For many families, eczema, or atopic dermatitis, begins in infancy with a patch of dry, red, itchy skin on the cheeks of a baby. It disrupts sleep, frustrates feeding, and quickly becomes the centre of daily life. But eczema doesn't always stay in childhood. It changes, adapts, and reappears – sometimes in adolescence, other times in adulthood, and even after years of calm.

What is eczema?

Eczema is a common, chronic inflammatory skin condition that affects around 10–30 per cent of children and up to 10 per cent of adults – and it's on the rise globally. Most cases begin early in life: around 70–90 per cent before age five, and nearly half within the first six months. While many children grow out of it, about 40 per cent continue to have symptoms into adulthood. For others, eczema starts later, often showing up on the hands, face, or eyelids.

Atopic eczema is part of a bigger story, sometimes called the atopic march. Babies with eczema are more likely to go on to develop food allergies, asthma, or hay fever. Skin with eczema is naturally dry and more permeable, making it easier for allergens, microbes, and irritants to sneak in and stir up inflammation. Genetics play

a role too – mutations in the filaggrin gene (FLG) affect the skin's ability to stay sealed and protected and allow more of that loss of water across the skin barrier.

Eczema triggers

Day-to-day triggers can make things worse – harsh soaps, central heating, teething, infections, stress, and certain foods can all set things off. Finding triggers can involve some detective work: changing washing powders, examining fabrics, and keeping food diaries.

Common irritants include soaps, wool, certain metals, preservatives in topical products, and house dust mites. Skin infections, particularly with *Staphylococcus aureus*, are common, while viral infections such as eczema herpeticum must be recognized and treated urgently.

How the condition modifies

Clinically, eczema changes with age. In babies, it often appears on the cheeks, scalp, and the back of the elbows or front of the knees. In toddlers, it shifts to the creases of elbows and knees. In older children and adults, the pattern may become more localized – commonly the eyelids, hands, and neck. With eczema, the skin is dry, itchy, and prone to cracking. Scratching leads to further inflammation, risk of infection, and often sleep disturbance. Over time, skin may become thickened as a result of chronic scratching.

Yet the burden of eczema goes beyond the skin. Children with eczema often experience disrupted sleep, poor concentration, and reduced quality of life, while parents can feel exhausted and helpless. Adults may struggle with appearance, discomfort, and the frustration of unexpected eczema flares, particularly when this will cause poor sleep and impact on their job and in communication with others.

Treatments for eczema

Managing eczema begins with restoring the skin barrier with emollients. Apply them generously, usually once a day, using products that combine humectants (to draw water into the skin), occlusives (to seal it in), and ceramides (to restore the lipid layer). Cleansers should be gentle and soap-free. It is important to find a moisturizer you like, and my advice is to put it on immediately after a lukewarm shower, before drying the skin.

Topical corticosteroids are used to control inflammation. When used correctly, they are safe and effective. Calcineurin inhibitors – which block enzymes in the inflammation pathway – are non-steroid alternatives for sensitive areas and long-term maintenance. Examples include pimecrolimus and tacrolimus.

Management of eczema follows a stepped approach:
- **Mild:** emollients + mild topical corticosteroids
- **Moderate:** emollients + moderate-potency corticosteroids + calcineurin inhibitors
- **Severe:** emollients + potent corticosteroids + calcineurin inhibitors + bandages, phototherapy, or systemic treatments as needed

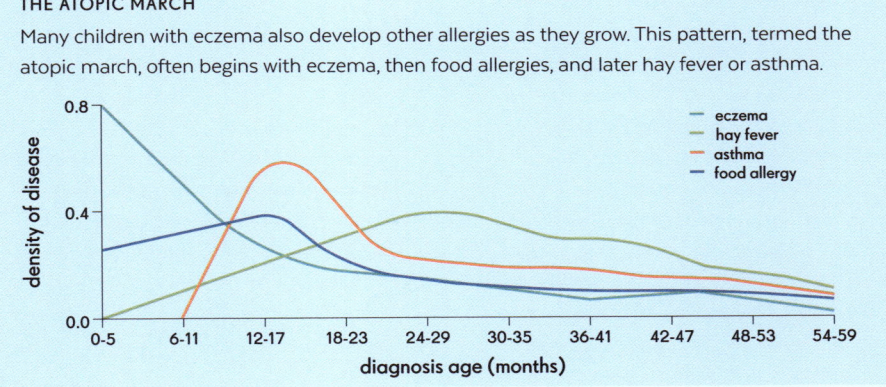

THE ATOPIC MARCH

Many children with eczema also develop other allergies as they grow. This pattern, termed the atopic march, often begins with eczema, then food allergies, and later hay fever or asthma.

Legend: eczema, hay fever, asthma, food allergy

y-axis: density of disease (0.0, 0.4, 0.8)
x-axis: diagnosis age (months) — 0-5, 6-11, 12-17, 18-23, 24-29, 30-35, 36-41, 42-47, 48-53, 54-59

psoriasis

Psoriasis is a skin condition that is both familiar and misunderstood. Many people have heard of it, often confusing it with eczema or dry skin, but few really grasp its complexity.

Psoriasis is a long-term skin condition driven by the immune system. At the centre of it all is a type of immune cell called a T-helper cell – specifically the Th17 variety – that becomes overactive and sparks a cycle of inflammation. Immune cells in the skin, like dendritic cells and macrophages, release chemical messengers called cytokines. These activate T-cells, which then go into overdrive. Th17 cells send out even more inflammatory signals, telling skin cells to grow and divide too quickly. The result is the thick, scaly patches we see on the skin – and in some cases, inflammation that also affects the joints.

Symptoms and causes of psoriasis

Psoriasis can start at any age, but it most often appears either in young adulthood or later in life, between 50 and 70. The most common type – plaque psoriasis – makes up 80 to 90 per cent of cases. These are the classic thick, salmon-pink patches covered in silvery scale. They tend to show up on the elbows, knees, scalp, shins, and lower back, though they can appear anywhere.

There are other forms too, including guttate psoriasis (small, teardrop spots that often follow an infection), pustular (painful, pus-filled bumps), erythrodermic (a rare, dangerous form with widespread redness and shedding), and inverse (which affects folds of the skin such as the armpits and buttocks).

Psoriasis isn't just a skin condition. Around a third of people also develop psoriatic arthritis, which causes joint pain and swelling, and can lead to permanent damage. Psoriasis is also linked to a higher risk of cardiovascular disease, type 2 diabetes, inflammatory bowel disease, and depression – reminding us that this condition goes far deeper than the surface.

How does psoriasis arise?

In people who are genetically susceptible – especially if both parents have it – triggers can switch on the condition. These include infections, stress, alcohol, smoking, and some medications such as lithium or beta-blockers. Not everyone with a family history will develop it, but when a particular mix of genes and environmental factors come together, the immune system overreacts and psoriasis begins to appear on the skin.

-)-)-)-

Treatment options

Every patient's psoriasis journey is different. Some need only topical therapy now and again while others live with constant flares and remissions.

For mild psoriasis, treatment usually starts with topical medications to calm inflammation and slow down skin cell growth. These treatments work best when used consistently and alongside emollients. Phototherapy (see pages 148–49) is used for those with more extensive disease.

In more severe cases, oral immunosuppressants are still used. The real revolution in psoriasis care has come from biologics (see pages 146–47) and, in well-selected patients, they can lead to near-complete clearance of symptoms and dramatic improvements in quality of life.

SITES OF THE DIFFERENT TYPES OF PSORIASIS

Generally, erythrodermic and pustular psoriasis can affect anywhere on the body, while plaque, guttate, and inverse psoriasis are more localized.

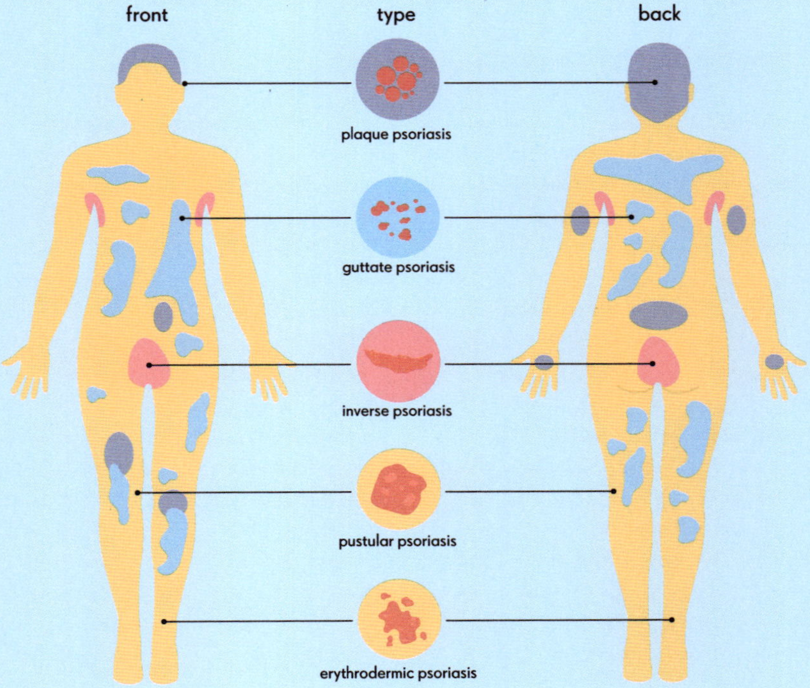

front

type

back

plaque psoriasis

guttate psoriasis

inverse psoriasis

pustular psoriasis

erythrodermic psoriasis

rosacea

Rosacea is a common, chronic skin condition that affects the face and sometimes the eyes. Let's examine the symptoms and causes of the condition, look at its triggers, and see what can be done to treat it.

Rosacea often creeps in slowly, sometimes mistaken for adult acne, sensitive skin, or even just a tendency to blush. It usually begins between the ages of 30 and 50 and is incredibly common, affecting around 1 in 10 people, though many don't realize that they have it.

How the condition presents

Rosacea is most often diagnosed in people with fair skin, particularly those of Celtic background, but it absolutely affects people with darker skin tones too. The difference is, the signs can be more subtle – less obvious redness means its diagnosis is often missed. Some people flush easily, with sudden waves of heat and redness across their face. For others, there's a more constant background redness, often over the cheeks, nose, chin, or forehead. Breakouts can also appear – red bumps or spots that look a bit like acne, but without blackheads and without the same oily skin.

In more advanced cases, the skin may start to thicken, particularly around the nose – a change known as rhinophyma. The eyes can be affected too, and may feel red, gritty, sore, or sensitive to light (a type called ocular rosacea).

Doctors have traditionally grouped rosacea into four main types, depending on which features are most prominent: flushing and visible vessels, spots and bumps, eye symptoms (ocular rosacea), or thickened skin (phymatous rosacea. In reality, most people don't fit neatly into one category – the signs and symptoms often overlap and can change over time.

What causes rosacea?

Rosacea is driven by a mix of immune, microbial, nerve, and blood-vessel factors – and is a story of overreaction. It fuels inflammation and makes blood vessels more visible, leaving the skin red, sensitive, and easily triggered.

One of the more unusual characters in this story is *Demodex folliculorum* – a tiny mite that lives on all human skin, mostly inside hair follicles and oil glands. We all have them, but in rosacea-prone skin, there tends to be more of them – and more reactivity to them. It's not just the mites themselves that are the issue; it's the bacteria they carry. One in particular, *Bacillus oleronius*, seems to stir up the immune system when the mites die and release it into the skin. This might help explain why some people

-)-)-)-
 ·.·.·

experience a flare at the start of treatment – as the mites die off, they release bacteria that briefly make the inflammation worse before it gets better.

Basic treatment

Start by keeping a flare diary. Not everyone reacts to the same things, and guessing often causes more stress than clarity. For a few weeks, jot down what you eat and drink, what skincare you use, how you feel, and what the weather's doing. Then note how your skin responds. The diary doesn't need to be perfect – just enough to spot patterns. This can help you identify your own triggers and adapt, rather than avoiding everything "just in case" (yes, you might still be able to enjoy coffee!).

   Stick to a simple, consistent skincare routine. Rosacea-prone skin tends to have a weakened barrier, which makes it more reactive to things such as water, temperature changes, and certain ingredients. The rules here are straightforward: cleanse gently (no foaming or stripping products), use a barrier-repair moisturizer, and apply a broad-spectrum SPF every morning. Once your skin settles, you can explore treatments – but this is always the foundation.

Medical treatment

For background redness and flushing:
- Topical vasoconstrictors like brimonidine or oxymetazoline
- Oral beta-blockers or clonidine for severe flushing
- Vascular lasers for visible vessels

For papules and pustules:
- Topical ivermectin, azelaic acid, or metronidazole
- Low-dose oral tetracycline antibiotics

For ocular rosacea:
- Warm compresses, lid hygiene, lubricating drops
- Oral tetracyclines
- In some cases, ciclosporin eye drops

For phymatous rosacea:
- CO_2 laser or surgical options to reduce tissue overgrowth

• ROSACEA TRIGGERS •

Rosacea triggers vary, but often include sun, heat, alcohol, spicy food, stress, exercise, hot drinks, and harsh skincare.

-)-)-)-

melasma

This condition often appears quietly as a subtle darkening
across the cheeks, upper lip, or forehead. It is often triggered
by changes in hormones and the affected area only requires
tiny amounts of UV light to get darker.

Melasma is a chronic, relapsing disorder of pigmentation that predominantly affects women between the ages of 20 and 40. Often triggered by pregnancy, it's most common on the face, though some cases extends to the forearms and upper back. Melasma is especially prevalent among people of Latin American, South-East Asian, and South Asian backgrounds, with studies reporting rates as high as 40 per cent. Though harmless, its location and persistence give it outsized impact – patients describe feelings of embarrassment, reduced self-esteem, and limitations in social and professional life.

Causes and presentation of melasma

The precise cause of melasma remains elusive, but usually a melanocyte (see page 14) becomes overactive due to genetics, hormonal stimulation like the oral contraceptive pill or pregnancy, or topical inflammatory reactions. Once that melanocyte is overactive then even tiny amounts of stimulation cause it to make pigment. This stimulation is predominantly caused by ultraviolet radiation (see pages 94–95) – particularly UVA – but visible light and

infrared radiation also play a more minor role. Clinically, melasma presents as symmetrical light to dark brown patches with irregular borders, typically in one of four patterns:

• **Centrofacial:** forehead, cheeks, upper lip, nose (most common)
• **Malar:** cheeks and nose
• **Mandibular:** jawline and chin
• **Extrafacial:** forearms and upper chest (more common in postmenopausal women)

Managing the condition

There is no cure for melasma, but there are many ways to manage it.

01. Treatment begins with understanding the chronic nature of the condition and the importance of daily sun protection.

02. Broad-spectrum sunscreen with SPF50+ is key. It needs to have excellent coverage in the UVA range (see pages 94–95). I tell my patients to apply this to their face and then apply another layer – every morning, every day of the year. A good top-up after lunch is key on long summer days. If using make-up, then a foundation containing iron oxide should be sought out for additional protection. Also, find a hat you love to wear, especially in the summer months, and learn to love the shade! A gentle skincare routine and cosmetic camouflage can be transformative.

03. With topical medical treatments, first-line therapies target melanin production.

- Hydroquinone and tretinoin remain the gold standard. They work together to reduce pigmentation and enhance skin cell turnover
- Azelaic acid is well-tolerated, particularly in epidermal melasma
- Kojic acid is less reliable and can cause irritation
- Cysteamine, niacinamide, and vitamin C are useful for maintenance or additional therapy

04. Oral tranexamic acid is emerging as a safe and effective second-line therapy, especially in cases resistant to topicals, but must be prescribed with caution in individuals at risk of thrombosis.

The myth of laser treatment and peels

Many people think that lasers (see pages 155–56) and peels (see page 151) are the go-to treatments for melasma – THEY ARE NOT! While they can be helpful at removing pigment quickly, if not used in conjunction with the correct topical treatment and with strict UV protection, only a third of patients get better, a third experience no difference, and the condition actually gets worse for the final third. If these lasers and peels are used, then topical treatment and SPF is key.

Melasma responds slowly, and visible improvements may take weeks, even months; relapses are frequent, particularly in summer or with hormonal changes. Once the condition is under control, continuing to apply sun protection is essential or the melasma will just come back.

Tiny amounts of UV light can trigger melasma, so strict application of SPF is essential.

-)-)-)-

vitiligo

Vitiligo is a long-term skin condition where white patches
appear due to the loss of pigment-producing cells in the
skin, called melanocytes. While vitiligo can affect anyone,
it's often more noticeable in people with darker skin tones.

Vitiligo affects around 1 in 100 people globally.
Its white patches develop when the immune
system becomes overactive and mistakenly
attacks melanocytes (see page 14). This
immune misfire is believed to be triggered by
a combination of genetics, skin injury, emotional
stress, or even sunburn. A key immune pathway
involved is the interferon gamma signalling loop,
which drives inflammation and, over time,
pigment loss in affected areas.

Types and treatments

There are two main types of vitiligo. Non-
segmental, or generalized, vitiligo, the most
common form, tends to appear symmetrically
on both hands, around the eyes or mouth, or
on the knees and elbows. It can start at any age
and often progresses slowly. This type is often
associated with other autoimmune conditions,
such as thyroid disease, type 1 diabetes, or
alopecia areata. Segmental vitiligo, in contrast,
usually affects only one area of the body,

appears earlier in life, and is less likely to
be linked to other medical issues.

While there's currently no cure, treatments
aim to reduce inflammation and encourage
repigmentation. These include topical
corticosteroids, calcineurin inhibitors,
narrowband UVB phototherapy, and newer
options such as ruxolitinib cream. Early
treatment, particularly on facial skin, can
produce better results. Patches on the hands
and feet tend to be more resistant. There are
surgical treatment options for vitiligo as well,
where functioning melanocytes are transplanted
to areas of vitiligo in an attempt to repopulate
the skin and regain pigment.

Why it matters emotionally

The psychological impact of vitiligo can be
significant and some people experience
embarrassment, anxiety, or depression,
particularly if their patches are visible. But vitiligo
isn't always seen as something to be treated

-)-)-)-

VITILIGO LOCATIONS ON THE BODY

The different types of vitiligo can present from the top of the head to the toes.

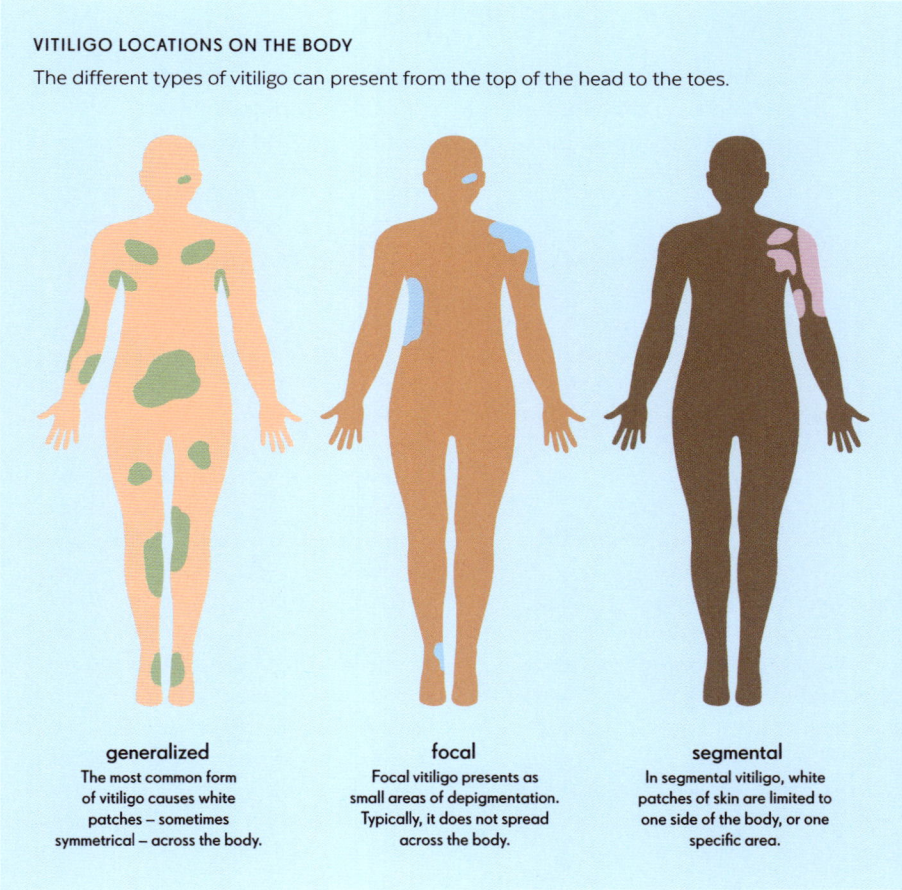

generalized
The most common form of vitiligo causes white patches – sometimes symmetrical – across the body.

focal
Focal vitiligo presents as small areas of depigmentation. Typically, it does not spread across the body.

segmental
In segmental vitiligo, white patches of skin are limited to one side of the body, or one specific area.

or hidden. Increasingly, in the media people are celebrating their skin – from models on magazine covers to everyday individuals proudly sharing their journeys online. For some, treatment is about regaining pigment while for others, it's about confidence and acceptance.

Either way, the most important thing is that everyone with vitiligo feels supported, respected, and empowered to choose the path that's right for them and given access to treatment when wanted.

-)-)-)-

allergies

Skin allergies arise when your immune system mistakes
something harmless for a threat and goes into defence
mode. It's trying to protect you, but it's got the wrong idea.

There are a few different ways allergies can show up on the skin. Some reactions are immediate, others take a day or two to appear. In all cases, your body is responding to something it sees as dangerous, even if it isn't. Understanding the type of reaction is the first step towards finding relief and avoiding the trigger in future.

Allergic contact dermatitis

This is the most classic form of skin allergy. It's what is known as a type IV hypersensitivity reaction, a delayed immune response that usually shows up 24–72 hours after exposure to a substance. On first contact, your skin doesn't react, but your immune system quietly becomes sensitized – with future multiple exposures, memory cells kick in and mount an inflammatory attack.

The rash usually appears as a red, itchy, well-defined patch where the allergen touched the skin, but usually it spreads beyond the original area. Over time, skin can become thickened, dry, or cracked – particularly with repeated exposure.

Common culprits are:
- Nickel (in jewellery, bra clips, belt buckles, phones)
- Fragrances (perfumes, detergents, even "natural" oils)
- Preservatives (like methylisothiazolinone [MI]) and methylchloroisothiazolinone [MCI] found in multiple skincare products)
- Hair dye (especially para-phenylenediamine [PPD]), also found in henna tattoos
- Rubber chemicals (gloves, elastic, adhesives)
- Plants (chrysanthemums, tulips)

Sometimes, figuring out what's causing a skin reaction feels a bit like detective work. We ask about your job, hobbies, skincare, even the shampoo you use; anything that might hold a clue. You can start your own investigation at home, but the most reliable way to get answers is with patch testing. This involves applying tiny amounts of common allergens to the skin, usually on your back, and checking for reactions over the next couple of days. If something triggers a response, we've found the culprit. From there, it's all about avoidance. That can be tricky, but

-)-)-)-
·:·:·:·

most products list their ingredients on the label or online, and once you know what to look for, you're already one step ahead.

Urticaria

Urticaria, also known as hives, is a different reaction known as a type I hypersensitivity – meaning it happens fast, often within minutes to hours of exposure. You'll usually see red, raised, incredibly itchy welts that appear suddenly, then vanish just as quickly, sometimes without a trace. In some cases, the reaction causes swelling around the eyes, lips, or even the throat. This is called angioedema, and while it often settles on its own, it can occasionally be serious.

Triggers include:
• Certain foods (nuts, shellfish, eggs)
• Medications (antibiotics, painkillers)
• Insect bites
• Cold, heat, pressure, exercise, or stress

Sometimes, we never find the cause. This is called chronic spontaneous urticaria (see box below), and it can last for months or even years; often this may be triggered by a viral infection.

Treatment usually begins with regular use of non-drowsy antihistamines. These are the same medications you might take for hay fever, but for hives, we often need to dial up the dose, under medical guidance. The aim is to calm the immune system just enough to stop it from reacting so dramatically.

If the hives persist after this initial treatment, there are other options. One of the most effective is a biologic called omalizumab. It works by targeting a key player in the allergic response – immunoglobulin E (IgE) – and helps quieten the immune system's overreaction. For people with chronic urticaria that hasn't responded to antihistamines, this treatment can be life-changing.

• CHRONIC URTICARIA •

Not all hives are caused by an allergy. Chronic spontaneous urticaria is an autoimmune condition where the immune system mistakenly activates the mast cells (see pages 16–17). Chronic inducible urticaria is triggered by specific stimuli such as cold, heat, and water.

-)-)-)-

scalp and hair disorders

Receding hairlines, widening partings, patchy bald spots, dry areas, or just more hair in the shower drain than usual could mean you have a scalp condition. Understanding what's causing it is the first step to doing something about it.

Like our face, our scalp and hair are visible to others, so changes here can often be a source of worry for many people, especially those involving hair loss and thinning. Let's take a look at some of the disorders.

Non-scarring hair loss

This is the most common group of hair disorders. The good news is that the hair follicles are still intact, so regrowth is often possible. Androgenetic alopecia, also called male or female pattern hair loss, usually comes with advancing age due to changes in circulating hormones. It is driven by sensitivity to DHT, a by-product of testosterone. It causes gradual shrinking of hair follicles. In men, it tends to show as a receding hairline and thinning at the crown. In women, it usually appears as a widening parting and general thinning over the crown, while the hairline stays put. Treatment varies depending on the type and severity, but can include topical treatments or oral agents and in some cases hair transplantation.

Alopecia areata is a different type of hair loss, caused by the immune system mistakenly attacking hair follicles. Under the microscope, this looks like a "swarm of bees" as white cells cluster around the follicle. It often starts with smooth, circular bald patches on the scalp or beard. Treatment for alopecia areata often includes topical or injected steroids, with newer options such as JAK inhibitors showing promise in more extensive or stubborn cases.

If all scalp hair is lost, it's called alopecia totalis. If all body hair goes, it's alopecia universalis – this can affect more than just appearance, making the eyes, nose, and skin more vulnerable to dust and allergens. A classic sign is the presence of "exclamation mark" hairs – short, broken strands that taper at the base. The course can be unpredictable. Hair might grow back on its own, or the condition may persist or progress. Other autoimmune conditions are often present too.

Scarring alopecia

Scarring (or "cicatricial") alopecias are less common but more serious. In these conditions, inflammation permanently destroys the hair follicle and replaces it with scar tissue, making regrowth impossible, so early diagnosis is vital.

Frontal fibrosing alopecia (FFA), a subtype of lichen planopilaris (LPP), causes a slowly receding hairline, especially in postmenopausal women.

It often starts at the very front of the scalp with a faint red ring around each follicle. Eyebrow loss is also common, with the same dotted redness. LPP itself can affect any area of the scalp and often comes with burning, itching, or tenderness. It also causes inflammation around follicles.

Lupus, particularly the discoid form, may also lead to scarring. It is marked by red, scaly, well-defined plaques on the scalp.

The key to all types of scarring hair loss is early recognition and prompt treatment, usually with anti-inflammatory medications to stop further damage.

Flaky scalp

A flaky, itchy scalp is incredibly common – but it's not always due to dryness. In seborrhoeic dermatitis, for example, greasy, yellowish scale is triggered by yeast on the scalp. This is what we often call dandruff (see page 162), and it

tends to flare with stress or cold weather. Scalp psoriasis causes thicker, silvery scales with a red base. It may extend beyond the hairline and be tightly stuck down, making treatment difficult. Other causes include contact dermatitis from hair dye or shampoo, and even fungal infections.

Medicated shampoos with ingredients like ketoconazole, coal tar, or salicylic acid can help. Severe cases may need topical steroids or prescription anti-inflammatories.

Telogen effluvium

If you've noticed handfuls of hair falling out a few months after a stressful event – surgery, childbirth, illness, or crash dieting – it could be telogen effluvium. This temporary hair loss happens when many hairs shift into the resting (telogen) phase at once. The good news? The follicles aren't damaged, and hair usually regrows on its own within 6 to 12 months.

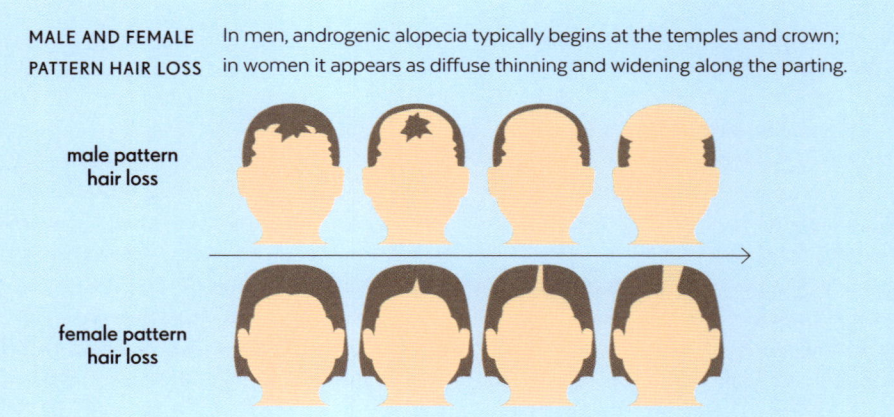

MALE AND FEMALE PATTERN HAIR LOSS In men, androgenic alopecia typically begins at the temples and crown; in women it appears as diffuse thinning and widening along the parting.

male pattern hair loss

female pattern hair loss

skin cancers

Skin cancer is the most common cancer in the world, and the numbers are still rising. That can sound worrying, but the good news is that most types are highly treatable if caught early. Simply being aware of your skin, and knowing what to look for, can make all the difference.

As dermatologists, we divide skin cancer into two broad categories: melanoma and non-melanoma. They behave quite differently, but the root cause is often the same – years of ultraviolet (UV) exposure gradually damaging the DNA in skin cells. This damage builds up silently, long before anything appears on the surface.

Melanoma

Melanoma develops in the pigment-producing cells of the skin, called melanocytes. It's the most serious form of skin cancer because, unlike many others, it can spread to other parts of the body. But again, early detection changes everything.

You might imagine melanoma as a large, obviously dark mole; most of them are, but not always. Some are pink, skin-coloured, or even colourless. One of the most helpful clues is the "ugly duckling", a mole that looks or behaves differently from the others on your body. The most important thing to notice is change.

If something on your skin starts to evolve – whether it's growing, itching, bleeding, or simply not healing – it's worth getting it looked at. Most of the time, it's nothing serious, but sometimes,

that small decision to seek advice is the thing that catches a melanoma early.

For moles, we use the ABCDE checklist:
- Asymmetry
- Border irregularity
- Colour variation
- Diameter over 6 mm (0.24 in)
- Evolution (any change over time)

Treatment usually involves removing the melanoma with surgery, often followed by a second, wider excision to be sure all cancer cells are gone. In some cases, nearby lymph nodes are checked too.

Lentigo maligna

Lentigo maligna is a type of melanoma in situ, meaning it's confined to the top layer of the skin and is not invasive. The risk of transformation to melanoma is low – studies suggest it's under 5 per cent in people over 75. When deciding what to do, we also have to consider the size and location. Surgery on a large patch, especially on the face or scalp, may carry its own risks

-)-)-)-

and complexities. In some cases, treatment might mean removal; in others, a topical immunotherapy like imiquimod, radiotherapy, or careful monitoring may be more appropriate.

Non-melanoma skin cancers

BASAL CELL CARCINOMA (BCC)

Basal cell carcinoma (BCC) is the most common skin cancer we see. It often shows up as a shiny, pink bump with a pearly edge that may bleed, then heal, then bleed again. It used to be called a rodent ulcer because its raised edge and hollowed centre resembled a rodent bite. BCC grows slowly and rarely spreads, but it can become harmful if left alone.

In high-risk areas like the nose, eyelids, or ears, Mohs surgery (where thin layers of skin are removed) offers the best chance of taking out the cancer while preserving healthy tissue. Simpler BCCs can often be treated with standard surgery, freezing, radiotherapy, topical creams, or even light-based therapy. The right approach depends on the type and location.

SQUAMOUS CELL CARCINOMA (SCC)

Squamous cell carcinomas (SCCs) can be more unpredictable. They might appear as a crusted lump, scaly patch, or a sore that bleeds or slowly grows. Early SCCs often behave like BCCs and stay local, but some can spread – especially those on high-risk areas like the lips, ears, or genitals, in immunosuppressed patients, or if they grow deeply. Treatment is usually surgical, with Mohs or radiotherapy used for more complex cases.

Before an SCC appears, the skin often gives us early clues – subtle signs of sun damage that haven't yet turned invasive. Actinic keratoses (AKs) are one of the most common. These feel like rough, sandpapery patches and tend to show up on areas that get a lot of sun.

Bowen's disease is another early warning sign. It's a form of SCC that's sitting on the surface and hasn't grown deeper. It usually looks like a red, flaky, or crusty patch. Both BCC and SCC are treatable with cryotherapy, creams, or minor procedures before they can become more serious.

SKIN MELANOMA CHECKLIST

The ABCDE checklist is a standard method of identifying potentially concerning moles.

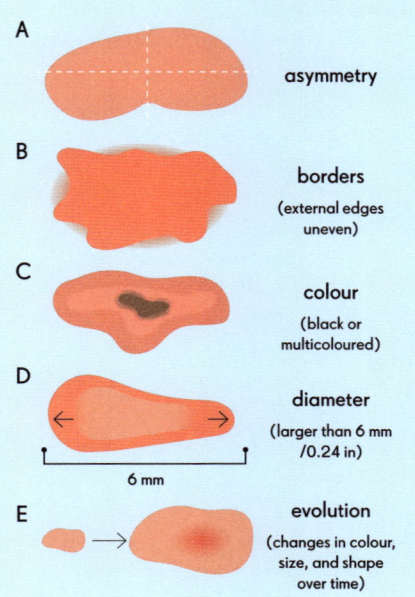

A — **asymmetry**

B — **borders**
(external edges uneven)

C — **colour**
(black or multicoloured)

D — **diameter**
(larger than 6 mm /0.24 in)

6 mm

E — **evolution**
(changes in colour, size, and shape over time)

infections

Skin is a brilliant barrier, but sometimes fungi, bacteria, or viruses manage to sneak past its defences – especially if there's a cut, weakened immunity, or an underlying condition such as eczema.

Fungal infections

Fungal infections – called dermatophytosis or tinea – tend to show up in warm, damp areas like the groin, feet, under the breasts, or between the toes. They often appear as red, scaly rashes with a ring-like shape and a clearer centre. Fungal nail infections can make nails yellow, thickened, and crumbly. Yeasts, like Candida, are another group of fungi that thrive in skin folds. They're more common in babies (as nappy rash), and in adults who are overweight or have diabetes.

Most fungal infections respond to antifungal creams like clotrimazole. If they're widespread or stubborn, oral medication may be needed.

Bacterial infections

- **Impetigo** is a superficial infection, often seen in children, causing honey-coloured crusts around the mouth or nose. It spreads easily.

- **Folliculitis** is an infection of hair follicles, seen as red, tender bumps.

- **Cellulitis** is a deep infection, causing hot, red skin (usually on the legs), often with fever.

- **Erysipelas** is a superficial skin infection often seen on the face. It requires antibiotics.

Bacterial infections are treated with antibiotics – creams for mild cases or oral antibiotics for deeper infections. If the area becomes very swollen, painful, or systemic symptoms develop, it's important to seek medical attention quickly.

Viral infections

- **Cold sores (herpes simplex)** are painful blisters, usually around the lips, caused by a virus that stays dormant in the body and reactivates.

- **Warts** are caused by the human papillomavirus (HPV), and appear as rough, thickened bumps on the skin, often on hands, feet, knees, and elbows. They can also occur in the genital area as a sexually transmitted disease.

- **Shingles (herpes zoster)** is a reactivation of the chickenpox virus, leading to a painful, blistering rash on one side of the body.

Antiviral drugs may help cold sores or shingles if started early. Persistent warts can be frozen, treated with salicylic acid, or removed in clinic.

benign lumps and bumps

Not every lump or bump is a cause for concern. In fact, most are harmless, though they can be itchy, annoying, or visible to others so become a source of embarrassment and affect quality of life.

Common benign skin growths

- **Seborrhoeic keratoses:** waxy, rough, or stuck-on-looking patches, often brown, grey, or black. They become more common with age and are often found at sites of friction, like bra-lines or waist-straps. They can be gently removed with freezing or scraping, laser, and electrosurgery.

- **Skin tags (acrochordons):** soft, dangling bits of extra skin found around the neck, underarms, eyelids, or groin. Often linked to friction, weight gain, or insulin resistance. Easy to snip or freeze off in clinic if they catch or bother you.

- **Epidermoid cysts:** round, firm bumps under the skin with a balloon-like cell wall and filled with dead skin cells – keratin (not pus). Usually harmless, they can have an impact on quality of life. If they become red, swollen, or infected, they may need draining or removal.

- **Lipomas:** soft, squidgy fatty lumps under the skin, often on the arms, back, or thighs. They move a little when pressed and grow slowly. No treatment is needed unless they become

uncomfortable or are growing, in which case, surgical removal is simple. They can get large and have an impact on quality of life.

- **Dermatofibromas:** firm, button-like nodules (often on the legs) that dimple when pinched. Usually develop after insect bites or trauma. Harmless and don't need treatment, though they can be itchy.

- **Campbell de Morgan spots (cherry angiomas):** bright red or purple dots on the torso that can appear in clusters as we age. These are tiny collections of blood vessels, and are completely harmless and very common. They can be lasered off if they bleed or for cosmetic reasons.

> ## • WHEN TO GET CHECKED •
>
> If a lump changes shape, grows quickly, bleeds, or just looks or feels different, always have it checked. Most are benign, but a quick dermatology review can give you peace of mind.

-)-)-)-

scars

Scars are the skin's way of repairing after injury, but unlike the original tissue, scar tissue is laid down quickly, more densely, and in a different pattern. You will also notice that there are no sweat glands and no hair follicles, which means it often looks and feels different.

Scarring can follow anything that injures the deeper layers of the skin – trauma, surgery, infection, inflammation, or burns. Scars come in a few different forms.

Atrophic scars are sunken or pitted and often follow acne or chickenpox. They're caused by damage to collagen and a loss of tissue. Hypertrophic scars are raised and thickened but stay within the original wound's borders, common after burns or surgery. Keloid scars go further – they grow beyond the original injury, forming shiny, sometimes painful or itchy overgrowths. In some people, even a tiny scratch can result in a large keloid; this is known as keloidal tendency.

Unborn babies before 24 weeks gestation can heal without scarring. Foetal skin can regenerate completely, thanks to a calmer immune response, a different collagen structure, and an environment rich in growth factors. After that point, skin starts to heal more like ours – leaving scars behind.

Acne scarring

Acne scarring deserves special attention because it is so common and visible. Types include:
• Ice pick scars (deep and narrow)
• Boxcar scars (wider with sharp edges)
• Rolling scars (soft, undulating depressions)
• Post-inflammatory erythema or pigmentation, which may fade with time but is often mistaken for true scarring

Treating scars

The goal with scar treatment is never total erasure – that simply isn't possible. Instead, it's about softening, fading, and improving texture. SPF use (see pages 126–27) is one of the most important things you can do, especially in the early stages of treatment. Silicone gels or sheets remain a go-to for early, hypertrophic surgical scars, and techniques such as massage and pressure garments can be useful in some cases.

The key to best treatment is understanding the type of scar. Laser therapies (see pages 155–56) are a mainstay and can help with redness, texture, and thickness. A special mention to fractional CO_2 lasers as these have been a game-changer for burn scars – they release tight, pulled areas of skin overlying joints to improve movement.

In addition, surgical techniques can soften particularly hard scars, while injectable treatments such as dermal fillers (see page 154) are used to lift dipped scars.

-)-)-)-

rare diseases

Rare skin diseases may affect only a few people, but they play a big role in dermatology. Many of these uncommon conditions are caused by changes in a single gene or proteins that are either defective or missing.

Studying rare skin conditions has helped us unravel not just how the skin functions, but also how DNA repair, cell adhesion, and immune regulation work across the entire body.

Some conditions, such as autosomal recessive disorders, only appear when a person inherits two faulty copies of the same gene – one from each parent. A single working copy is usually enough, but when both are "knocked out", the body can't perform a vital function, and disease results.

Examples of autosomal recessive disorders

People with xeroderma pigmentosum (XP) have a fault in one of the genes responsible for repairing DNA damage caused by ultraviolet (UV) light. This leads to severe sunburn, freckling, and an increased risk of skin cancers. Understanding XP has helped scientists learn how healthy skin detects and fixes UV damage.

Epidermolysis bullosa (EB) is a group of rare conditions where the skin is very fragile due to faulty or missing proteins. In severe forms, even light rubbing can cause painful blisters. Studying EB has revealed the importance of proteins in skin integrity, and led to advances in wound-healing.

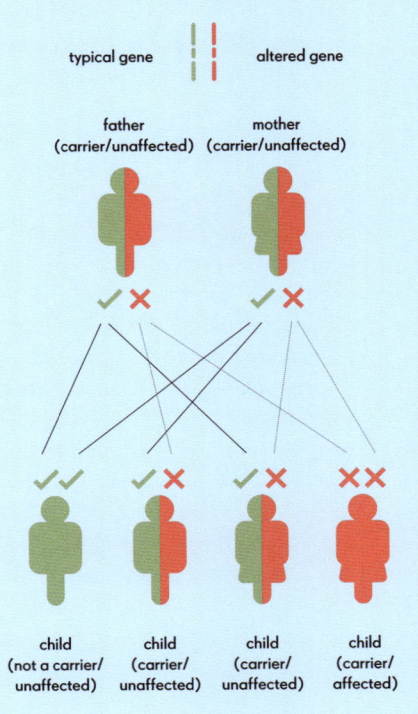

THE PATTERN OF AN AUTOSOMAL RECESSIVE DISORDER

If both parents are unaffected carriers, one of their four children will likely be affected.

typical gene   altered gene

father
(carrier/unaffected)

mother
(carrier/unaffected)

child
(not a carrier/
unaffected)

child
(carrier/
unaffected)

child
(carrier/
unaffected)

child
(carrier/
affected)

DOES EATING CHOCOLATE GIVE ME ACNE?

When we talk about food and pimples, the best evidence points less at cocoa itself and more at what tends to come with it, so things such as sugar, milk, and highly processed ingredients. High-glycaemic foods (those that spike blood sugar quickly) and frequent sweet snacks push insulin and IGF-1 (insulin-like growth factor 1) up, which encourages sebaceous glands to produce more oil. A standard milk chocolate bar is essentially a package of sugar plus dairy and fat. If your skin is already acne-prone, this combination might make your breakouts a bit more frequent or more inflamed. But don't blame the cocoa! Pure dark chocolate is a different story – it contains far less sugar and no milk, and the cocoa itself is rich in polyphenols that provide antioxidant and anti-inflammatory benefits. So, the honest answer is: chocolate on its own is very unlikely to be the single cause of your acne. Switch to darker, less sweet chocolate, and pay attention to the overall pattern of your diet rather than demonizing one food.

IS IT BAD TO POP PIMPLES?

I completely understand the urge – it may even be a factor as to why I wanted to become a dermatologist! Yet while popping pimples is very satisfying, it does more harm than good. A spot is essentially a tiny, inflamed, walled-off pocket in the hair follicle. When you squeeze it, you are not just pushing material outwards, you are also forcing some of that mixture of oil, dead cells, and bacteria sideways and downwards into the surrounding tissue. That can turn a small, superficial bump that would have settled in a few days into a deeper, angrier lump that hangs around for weeks, leaves a mark, or even becomes a true cyst. If you cannot resist tackling an obvious whitehead, the least damaging approach is to wait until there is a clear, soft white, or yellow head very close to the surface. Then cleanse your hands and the area, apply the gentlest pressure with cotton buds either side, and stop as soon as clear fluid or blood appears; follow with a bland topical antiseptic. Avoid touching deep, painful, under-the-skin spots, repeated picking at the same area, and going after every little bump.

DOES STRESS MAKE MY ACNE WORSE?

When you are under pressure, your body does not distinguish between a looming deadline and a genuine threat, so stress hormones such as cortisol and adrenaline rise. Cortisol nudges your oil glands to be more active and can also disrupt how skin cells shed inside the hair follicle, making it easier for little plugs to form and start the acne lesion. Stress also tends to quietly chip away at all the routines that help keep acne under control. You sleep worse, reach for more sugary snacks, forget to use your treatments consistently, touch your skin more, and rub your face when you are thinking. None of these cause acne on their own, but together they can magnify it. When patients tell me that their skin always flares around exams, in busy work seasons, or after major emotional upheavals, it makes biological sense. The flip side is important, though – you don't get acne just because you are stressed, and you cannot cure acne just by relaxing.

——

WHAT IS THE BEST WAY TO TREAT MOLLUSCUM CONTAGIOSUM?

Molluscum contagiosum is a very common, usually harmless viral infection of the skin that presents as small, hard, white spots with a central dimple. Most cases resolve within 6–18 months, but it can rumble on for two or three years, especially if there are lesions or eczema in the background. That is why actively watching and caring for the surrounding skin is beneficial. I start to think about treatment when molluscum is painful, very itchy, getting infected, clustering around the genitals, or causing a lot of distress. There is no treatment that reliably clears every molluscum overnight; the virus and the host immune system do most of the heavy lifting, although there are some topical treatments now that assist the immune system in this job. Physical removal (gentle squeezing with a comedone extractor, or pricking the core out with a fine sterile needle) is the most immediately effective approach: each lesion that is properly evacuated is usually gone for good. The downsides are obvious: it hurts (even with numbing cream), it is time-consuming if there are many spots, and it can leave temporary marks or, rarely, little scars, especially on darker skin or in children who wriggle and scratch.

——

-)-)-)-

05

A skin-loving lifestyle

sunlight

One of the most powerful forces that shapes our skin is sunlight, which has numerous positive effects on us. For many, the skin takes "the hit" for these benefits – but there are ways we can help it.

We often think of sunlight as heat and light, but it's actually a broad spectrum of electromagnetic radiation, and its first interaction with us is through our skin. The ultraviolet (UV) portion of the sun's radiation is divided into UVB and UVA rays, and together they deliver a "double hit" on the skin. The former causes direct mutations, while the latter accelerates tissue breakdown and repair failure. Over decades, this contributes not only to increased cancer risk but also to the visible and functional decline of the skin.

The differences between UVB and UVA radiation

UVB light is short-wave, high-energy radiation that mostly affects the epidermis. It is responsible for sunburn and is directly absorbed by the DNA in skin cells, causing DNA damage known as pyrimidine dimers. These mutations trigger a cascade of events that, if not repaired, increase the risk of skin cancers.

UVA light, by contrast, has a longer wavelength and penetrates more deeply into the skin – down to the dermis, where collagen, elastin, and blood vessels live. Rather than damaging DNA directly, UVA generates oxidative stress, producing reactive oxygen species (unstable molecules containing oxygen) that in large numbers can damage cell structures. This leads to photo-ageing – put simply, it makes our skin look older. UVA activates enzymes that break down collagen and damage elastin fibres. The result is skin that loses its firmness and elasticity, leading to the appearance of wrinkles and the texture of the skin becoming leathery or thickened.

Our skin has evolved a remarkable capacity to sense and respond to sunlight.

-)-)-)-
·:·:·:·

The effects of UVA rays

UVA stimulates melanocytes (see page 14) in a way that leads to the development of lentigines or "sunspots". It can cause dilation and inflammation of the skin's tiny blood vessels, leading to visible telangiectasia (the fine red lines often seen across the cheeks or nose). In some people, this is compounded by solar elastosis – the accumulation of abnormal elastin in the dermis that gives the skin a yellowish, sallow hue.

Although UVA is less efficient at causing mutations than UVB, its ability to suppress the skin's immune defences and trigger oxidative damage means it still plays a role in skin cancer development. Unlike UVB, it passes through glass and is present all year round, even on cloudy days, so for many, there is benefit from wearing SPF all year round.

When UV light hits the skin, it triggers melanocytes to produce melanin, the pigment that darkens the skin and gives us a tan (see page 20). Melanin absorbs UV radiation and helps reduce further DNA damage in skin cells.

As a dermatologist, I often see the damage caused by the Sun, but I also want to remind you of the benefits of sun exposure.

THE DEPTH OF THE SUN'S RAYS
Sunlight contains many wavelengths, each penetrating the skin to a different depth. UVB is mostly absorbed by the epidermis, where it causes direct DNA damage and triggers vitamin D synthesis. UVA travels much further through into the dermis. Visible light (400–700 nm) and infrared (760–1400 nm) reach the deep dermis and even the hypodermis.

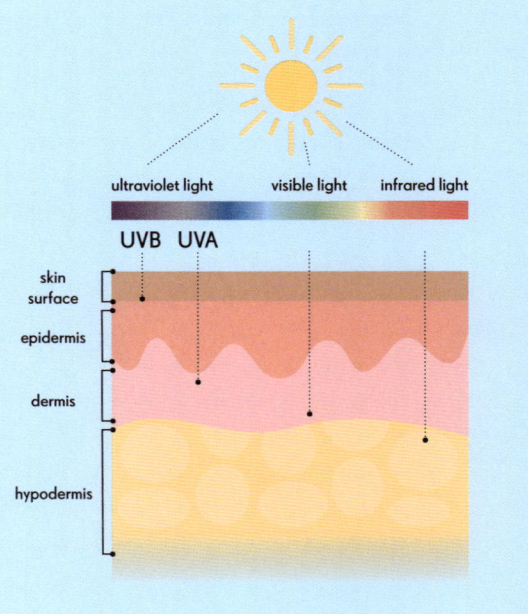

ultraviolet light   visible light   infrared light

UVB   UVA

skin surface

epidermis

dermis

hypodermis

Synthesis of vitamin D

When UVB light hits the skin – especially wavelengths around 295 to 297 nanometres (nm) – it triggers the conversion of a substance called 7-dehydrocholesterol into previtamin D_3. This then goes on to be processed by the liver and kidneys into its active form: vitamin D.

The skin is remarkably efficient at this. For many people, just short bursts of midday sun are enough to maintain healthy levels, and there's a clever built-in safeguard to stop overproduction. Once you've made enough pre-vitamin D_3, any extra UVB gets absorbed and diverted into other harmless, inactive compounds. This means vitamin D production naturally plateaus after about 10 to 30 minutes of strong sun, depending on your skin type and the intensity of the UV.

Lowering blood pressure

There's also emerging evidence that sunlight – particularly UVA – may play a role in regulating blood pressure. It appears to work by releasing nitric oxide stored in the skin, which helps widen blood vessels and lower systemic blood pressure. Some studies have even shown that people with more sun exposure tend to have lower rates of hypertension (high blood pressure).

In dermatology clinics, we harness UV light in controlled, measured doses to treat skin conditions like psoriasis and eczema, using its ability to calm the immune system. And of course, beyond the biological effects, there's something deeply human about being in the sun. We tend to move more, see people, and feel better – physically and emotionally – when we're out in the light and warmth.

Finding a balance

So how do we navigate the risks of UVA and UVB radiation and protect our skin best? It's really a balance, particularly to reduce the risk of skin cancers and to keep your skin looking as young as possible for as long as possible.

The goal isn't avoidance, it's management, and this can be achieved by following a few simple rules:

-)-)-)-

01. NEVER, EVER let your skin burn in the sunshine! This has no benefit to the body.

02. NEVER use tanning beds.

03. In the middle of summer, schedule outdoor activities early or late in the day to avoid the strongest UV rays between 11 am–3 pm.

04. Wear a wide-brimmed hat and UV protective sunglasses during peak sun intensity.

05. Apply a broad-spectrum SPF50 daily to the face and neck. This will reduce skin cancer risk in these areas. As an added bonus, the renowned Australian Nambour study showed that daily sunscreen use significantly slowed the development of wrinkles and other visible signs of ageing.

Sunscreen – choose wisely

The type of sunscreen you choose also matters. I like to think of it not just as sun protection, but as radiation protection, because what we're defending against isn't just heat or sunshine – it's the full spectrum of UV radiation, which is often present even when it doesn't feel sunny at all. For example, on a bright winter's day in London, UVA levels can still be high enough to drive oxidative stress in the skin to around one-third of midsummer levels. In addition, UVA passes through glass, which means your skin may be exposed even if you're indoors or in a car.

Sunscreen works as a genuine anti-ageing intervention – not in a cosmetic sense, but in a biological one.

-)-)-)-

nutrition

What you eat has a direct impact on how your skin looks, behaves, and repairs itself. Let's take a look at the nutrients and vitamins you need to ensure that your skin is as healthy as it can be.

The connection between diet and skin health often becomes clearest when something's missing. Take scurvy, for example, which is caused by a lack of vitamin C. Without the vitamin, collagen production slows, leading to easy bruising, poor wound-healing, and skin that loses its strength.

Zinc deficiency can show up as a scaly, inflamed rash, often around the mouth, in the folds of the skin, or near the perianal area. When the body lacks essential fatty acids, the skin can become dry, flaky, and irritated, sometimes looking a lot like eczema. Not getting enough protein slows the production of both keratin and collagen, making the skin more fragile and prone to injury.

Antioxidants and healthy fats

It's not just about preventing deficiency, it's also about supporting optimal skin function. Certain nutrients and food groups have been shown to provide benefits to skin beyond basic health.

The skin is a reflection of internal health, and while no single food will transform it, a consistent pattern of eating whole, nutrient-dense, anti-inflammatory foods can make a real difference in how your skin looks and functions over time. High-quality studies consistently support the advice that "eating a rainbow" of fruits and vegetables can slow skin ageing and promote a healthier complexion.

Diets rich in antioxidants that are found in colourful fruits and vegetables, green tea, nuts, and seeds help reduce oxidative stress in the skin and may protect against premature ageing. Carotenoids, the pigments that produce red, orange, and green colours, are particularly beneficial for skin appearance. They accumulate in the epidermis and form a protective shield against UV radiation (see pages 94–95) and oxidative damage. Compounds called polyphenols (including flavonoids and anthocyanins) found in berries, grapes, pomegranates, green tea, and other plant foods help combat inflammation and UV-induced oxidative stress in the skin. Clinical evidence suggests these compounds can improve skin quality when consumed regularly.

Healthy fats, particularly omega-3 fatty acids – with the most biologically active forms being EPA (eicosapentaenoic acid) and DHA (docosahexaenoic acid) from oily fish, flaxseeds, and walnuts – are incorporated into cell membranes in the skin, helping to reduce water loss and inflammation, and reinforcing the skin barrier.

-)-)-)-

The benefits of kimchee and green cabbage

Fermented foods – think kimchee and sauerkraut – and dietary fibre may support the gut microbiome (see pages 180–81), which increasingly appears to influence inflammatory skin conditions such as acne, eczema, and psoriasis.

Green leafy vegetables such as cabbage are rich in natural nitrate, a chemical compound the body converts into nitric oxide (NO), which relaxes blood vessels and improves circulation. This not only helps lower high blood pressure, but may also promote healthier skin by improving circulation and supporting the skin's barrier and immune functions.

Do I need to take supplements?

Supplements can, of course, be helpful in cases of deficiency or when dietary intake is limited, but they're not a substitute for a nourishing diet. There are some that require a mention. Omega-3 fatty acids supplementation may be especially helpful for those who don't eat oily fish regularly (three times per week), have dry or sensitive skin, or are managing chronic inflammatory dermatoses or medications that dry out the skin such as isotretinoin. Higher doses (typically 1–3 grams/0.035–0.1 oz per day of EPA and DHA combined) have shown benefit in improving symptoms and skin comfort, though lower daily amounts still support general skin health.

Vitamin D deserves a special mention. While sunlight is the main source, food-based and supplemental vitamin D can help maintain healthy levels during the winter months, particularly for those who see little UVB in that darker season.

Collagen supplements have become increasingly popular in recent years. These products typically contain hydrolyzed collagen peptides, which are broken down into amino acids and absorbed through the gut. While some studies suggest modest improvements in skin texture and hydration after 8 to 12 weeks of use, the evidence is far from conclusive. Most trials are small, industry-funded, and rely on subjective or short-term endpoints. I certainly don't spend my money on these.

Probiotics are live microorganisms that may help support gut health and, by extension, influence the skin. There's growing interest in their use for inflammatory skin conditions, but so far studies have been limited and strain-specific. While probiotics are generally safe, they're not a guaranteed solution for skin problems, and more high-quality research is needed before they can be confidently recommended for most people.

-)-)-)-

sleep and skin regeneration

Our skin follows a 24-hour circadian rhythm that regulates many of its functions. Daytime is geared toward skin defence and nighttime is prime for skin renewal, making good-quality sleep essential.

Nighttime is the skin's "restoration" phase. While we sleep, the skin barrier slightly relaxes, increasing transepidermal water loss (see page 18) and accelerating cell turnover and repair mechanisms. By morning, new cells have replaced some of yesterday's damage, though the skin may need rehydration due to moisture lost overnight. This natural rhythm means your skin spends the night focused on healing and the day protecting you all over again.

At night, the skin's outer barrier becomes more permeable, which explains why skin often feels driest by early morning. The flipside is that greater permeability at night can aid absorption of topical products applied before bed.

After daytime UV exposure, many DNA damage repair processes in skin cells peak during the night to fix mutations accrued by day. This includes nucleotide excision repair of UV-induced lesions, which has been found most active in the evening and night hours in humans.

Your skin does its deepest renewal work at night, with cell turnover reaching its peak around midnight. It's your body's way of making sure repair happens while you rest, saving energy for defence during the day.

The role of sleep-regulated hormones

Often called the "sleep hormone", melatonin surges at night, and plays a key role in the skin's renewal. Melatonin is a potent antioxidant that helps neutralize the day's oxidative stress, suppress UV damage in skin cells, and promote wound-healing in the skin. The nightly melatonin spike thus signals skin to enter "repair mode", enhancing DNA repair and bolstering the barrier. When sleep is disrupted (or in older individuals with blunted melatonin rhythms), the skin loses this protective, repair-boosting signal.

Cortisol, the body's main stress hormone, follows an opposite rhythm – its reduction each night is crucial for skin repair. At night, low cortisol allows collagen production and cell proliferation to proceed unhindered. However, the nighttime cortisol trough has a downside for inflammatory skin conditions, and many people with eczema or psoriasis experience worse nighttime itch and inflammation.

Deep sleep (especially slow-wave sleep earlier in the night) is when the pituitary gland releases growth hormone (GH), which is a major stimulus for tissue repair and keeping the skin strong.

The detrimental effects of poor sleep

Chronic poor sleep is linked to premature skin ageing. In a clinical study, women with poor sleep (5 hours or less per night) had significantly more fine lines, uneven pigmentation, and reduced elasticity compared to age-matched good sleepers.

Sleep deprivation is a pro-inflammatory state, and impaired immune cell function means wounds heal more slowly and the skin is more prone to irritation. Studies in sleep-deprived humans have noted that even small skin injuries take longer to heal when the subject is sleep-deprived.

A lack of sleep can trigger or aggravate common skin conditions, and the relationship is usually bidirectional. For example, acne itself can cause stress and insomnia, but it's clear that sleep deprivation promotes inflammation and hormonal imbalances that can trigger acne flare-ups. Other inflammatory conditions such as psoriasis or eczema may also flare with poor sleep. Yet, if sleep can be improved then so too can the severity of disease.

Good nighttime practices for better skin are:

01. For most adults, getting around 7–9 hours of sleep a night

02. Sleeping at night and in total darkness yields the most restorative sleep

03. Wearing breathable natural fabrics for pyjamas and bed linen

04. Using silk pillowcases to reduce friction on facial skin

General tips for a better sleep include maintaining a consistent bedtime, sleeping in as dark a room as possible, and avoiding screen light before bed.

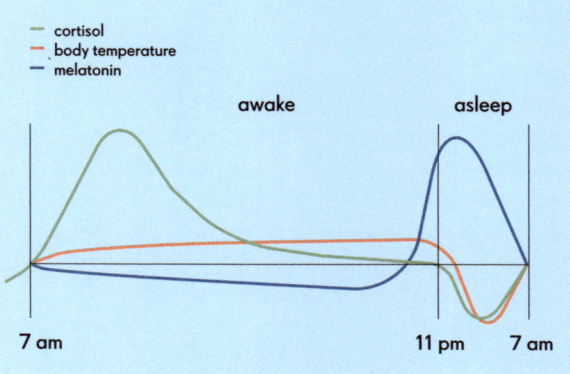

THE HORMONE CYCLE

Three key signals indicate a good night's sleep. Melatonin begins to climb a couple of hours before bedtime, stays high through the night, then drops sharply with morning light. Cortisol is lowest during deep sleep, then rises to help you wake up. Body temperature gently falls in sleep and reaches its lowest point around 3–4 am.

— cortisol
— body temperature
— melatonin

awake

asleep

7 am

11 pm   7 am

-)-)-)-

exercise

Increased blood to the skin, a boosted immune system to aid wound healing, and higher collagen levels are just some of the benefits of exercise. Let's examine how a workout can do wonders for your skin.

When you exercise, your heart pumps more blood around your body, including to your skin. This enhanced circulation brings more oxygen and nutrients to your skin cells and helps carry away waste products more efficiently. This is very relevant in the dermis, where collagen and elastin are produced.

What happens to the skin when you exercise?

During exercise, your muscles release small proteins called myokines. These travel through the bloodstream and send beneficial signals to other tissues, including the skin. They include messages to reduce chronic inflammation and promote mitochondrial (cell) health and collagen production in skin cells.

After each workout, there is an immune boost and a shift towards a healing and anti-inflammatory response, helping the skin repair without excess scarring or prolonged redness. In both animal and human studies, the wounds of people who exercise regularly heal faster than sedentary individuals.

Perhaps one of the best examples of this is the effect of exercise on ageing skin. In a study of older sedentary adults, participants who began a regular exercise programme showed remarkable changes in their skin after just 12 weeks. The stratum corneum, which had become thick and dry with age, returned to a more youthful structure. Even more impressively, the dermis, which naturally thins over time, grew thicker and showed higher levels of healthy collagen. These changes mirrored the qualities seen in much younger skin, suggesting that exercise actually reverses some of the skin's visible signs of ageing.

The effects of each exercise type

Different types of activity stimulate different responses in the body and skin.

ENDURANCE TRAINING: THE MITOCHONDRIAL BOOST

Long, steady-state activities such as running, cycling, and swimming are known as endurance exercises. These workouts challenge your cardiovascular system and boost the efficiency of your cells' mitochondria to improve the rate of natural age-related decline. Exercise also helps to keep the lymphatic system active, supporting the circulation of immune cells and the removal of cellular waste from tissues.

-)-)-)-

STRENGTH TRAINING: THE COLLAGEN CONNECTION

When you lift weights or engage in resistance training, your muscles exert short bursts of force that send a different set of signals to the body. These signals include hormones such as growth hormone (GH) and insulin-like growth factor 1 (IGF-1), both of which stimulate collagen production in the dermis. Collagen is the key structural protein that keeps skin firm, elastic, and wrinkle-resistant. With age, collagen levels naturally decline, but strength training can help slow that process.

In one trial involving middle-aged women, researchers compared the effects of aerobic versus resistance training over 16 weeks. Both groups saw improvements in skin elasticity, but skin biopsies from those in the resistance group revealed a thicker dermis. This suggests that strength training not only maintains but rebuilds the skin's structural support.

HIIT: CELLULAR YOUTHFULNESS IN A TIME-EFFICIENT PACKAGE

High-intensity interval training (HIIT) is a workout format that alternates short bursts of intense activity with periods of rest or lower-intensity movement. It increases the activity of an enzyme called telomerase, which protects the ends of our DNA strands (telomeres). Longer telomeres are associated with slower cellular ageing and greater regenerative capacity.

Hormone regulation

One of the ways exercise benefits the skin is by regulating hormones. Chronically high cortisol (the body's main stress hormone) can thin the skin, break down collagen, and impair barrier function. Regular physical activity has been shown to lower baseline cortisol levels and improve the body's stress response. Yoga and other mind-body exercises are especially effective at blunting cortisol spikes, and incorporating them into an exercise routine has been associated with improvements in inflammatory skin conditions like psoriasis.

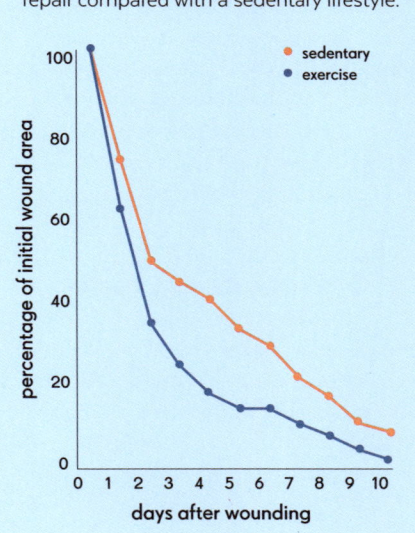

WOUND HEALING AFTER EXERCISE

Regular physical activity speeds up skin repair compared with a sedentary lifestyle.

percentage of initial wound area / days after wounding

- sedentary
- exercise

stress

Both short bursts of stress and long-term psychological pressure can have very real effects on our skin. This relationship between stress and the skin isn't one-way — our skin and brain are in constant conversation.

This close connection between skin and mind sits at the heart of a growing field called psychodermatology – where dermatology meets psychiatry and psychology. It explores how mental health can influence the skin, and how living with a skin condition can, in turn, affect our emotional wellbeing.

One of the key ideas in this area is the brain–skin axis – a two-way communication pathway linking our nervous system and skin. Emotional stress can trigger changes in the skin, and signals from the skin can influence brain activity. It's why a flare-up often appears during a stressful time, and why seeing that flare in the mirror can make us feel even more anxious. It's a feedback loop that can be difficult to break, but understanding it is the first step.

The stress-skin connection

When the brain registers stress, it activates a hormone cascade called the hypothalamic-pituitary-adrenal (HPA) axis. This prompts the release of cortisol (the classic "stress" hormone) and a set of messenger chemicals called catecholamines. These travel through the bloodstream and can trigger skin inflammation, recruit immune cells, or activate mast cells – the same cells involved in allergic reactions and itching.

But our skin isn't just a passive recipient – it can also produce its own stress hormones, responding to things such as pollution, extreme temperatures, or trauma. It sends chemical signals back to the brain, amplifying the stress response further. It's one reason why the skin, as the body's outermost barrier, plays such a pivotal role in our overall stress physiology. One of the lesser-known impacts of stress is its effect on the skin barrier itself by impairing barrier repair and increasing water loss from the skin, leaving it dry and prone to inflammation. In people with underlying skin conditions, this disruption can tip the balance and provoke a flare.

A common form of stress-related hair loss called telogen effluvium (see page 83) causes more hairs than usual to prematurely enter the shedding phase. This typically occurs a few months after a major life stressor like illness, surgery, bereavement, or psychological trauma.

-)-)-)-

Stress-related skin conditions

There are several specific conditions where psychological factors play a central role in the skin symptoms themselves. These include:

- **Trichotillomania**: an impulse control disorder in which a person pulls out their own hair, often unconsciously, as a response to stress or anxiety.

- **Dermatillomania**: repetitive picking at the skin, often at areas of perceived imperfections, leading to sores or scarring.

- **Delusional parasitosis**: a rare but severe condition in which individuals are convinced they are infested with parasites, despite medical evidence to the contrary.

These are not "made-up" conditions. They are real, distressing disorders at the crossroads of the brain and skin, and they require sensitive, integrated care.

And breathe…

While we can't eliminate all stress from our lives, there is growing evidence that mindfulness-based practices including meditation and deep breathing can lower cortisol and help rebalance the nervous system. In addition, many chronic skin conditions benefit from psychological support. Cognitive behavioural therapy (CBT) has been shown to improve outcomes in psoriasis and eczema. Mindfulness, relaxation therapy, and stress-reduction techniques are increasingly recognized as important complementary approaches to medical treatment.

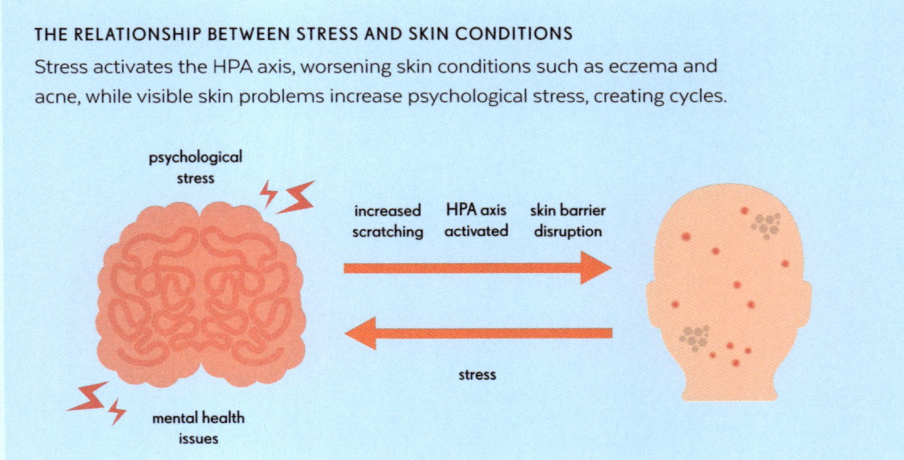

THE RELATIONSHIP BETWEEN STRESS AND SKIN CONDITIONS
Stress activates the HPA axis, worsening skin conditions such as eczema and acne, while visible skin problems increase psychological stress, creating cycles.

psychological stress

increased scratching · HPA axis activated · skin barrier disruption

stress

mental health issues

environmental factors

There are several factors in our environment that can
influence how our skin behaves, looks, and changes over time.
Let's take a closer look at the most consequential of them.

Air pollution

A significant environmental contributor to the
health of our skin is air pollution. We now know
that fine particulate matter found in air pollution
can sit on the skin and, in some cases, penetrate
through follicles or a compromised barrier.
Once inside, they generate oxidative stress
and contribute to inflammation, pigmentary
changes, and collagen breakdown.

This is particularly relevant in urban
environments, where long-term exposure has
been associated with accelerated photo-ageing,
even in those who protect themselves against
sunlight. It has even been shown that the skin of
dogs and cats living in environments with higher
levels of air pollution from car exhausts have
higher levels of pigmentation than their
non-urban counterparts.

Smoking

Tobacco smoke is another well-documented
potent inducer of oxidative stress. Smoking
reduces oxygen delivery to the skin, triggers
free-radical formation (see page 20), and alters
collagen and elastin synthesis; it also actively
stimulates enzymes that break down collagen.
Over time, this leads to a characteristic pattern
of deeper wrinkling, dullness, and a yellow-grey
tone to the skin – particularly around the mouth
and eyes. It also impairs wound-healing and
depletes key nutrients needed for skin repair.

The weather

The climate also has an impact on skin health.
Dry weather strips the skin of its natural lipids

and draws out moisture, while hot, humid conditions can increase sweating, which impacts inflammatory conditions such as intertrigo – a rash that appears in warm, moist areas where skin touches skin like under the breasts, in the groin, or between folds.

Wind exposure increases evaporation and can trigger sensitivity in rosacea-prone skin, while air-conditioning and central heating both lower ambient humidity and dry out the stratum corneum (see pages 18–19).

Other factors

Volatile organic compounds (VOCs) released from cleaning agents and fragrances from candles can act as irritants or allergens for some people. Heavy metals found in cigarette smoke and unregulated cosmetics can have the same detrimental effect on the skin.

What can you do?

There are several things you can do to mitigate damage from elements in our environment we know cause harm to your skin.

01. Reduce exposure to known factors – do not smoke tobacco products.

02. Cleanse the exposed skin properly at the end of the day.

03. Moisturize to ensure good hydration of the skin barrier.

04. Incorporate topical antioxidants into your skincare routine.

05. Support your internal antioxidant system with a colourful diet and good sleep.

Avoid scaremongering groups online claiming – without good evidence – that skincare ingredients are toxic.

-)-)-)-

DOES SUN PROTECTION FACTOR (SPF) DAMAGE CORAL REEFS?

In the lab, several common chemical filters found in SPF can harm corals and other marine organisms. At high enough concentrations, they have been shown to promote viral infections in symbiotic algae, trigger bleaching, damage DNA, and disrupt larval development. This is the scientific basis behind bans in places such as Hawaii and Palau, which restrict sunscreens containing certain filters in order to protect heavily visited reef areas. However, when you look at real-world concentrations in seawater, most monitoring studies find these UV filters at extremely low levels, far below those used in lab studies. The main drivers of reef damage are still warming seas, acidification, pollution, and overfishing, with sunscreen a comparatively small contributor. Quite probably, the CO_2 emissions generated by the aircraft that takes you to Hawaii may be more important for the water temperatures and the coral reefs than the SPF filters. My recommendation to patients is to wear their SPF on exposed sites when swimming, but to cover other parts of the body with protective swimwear.

———

DO I REALLY NEED TO WEAR AN SPF IN WINTER?

If you think about your skin in terms of lifelong wear and tear rather than single sunburn days, then yes – there is a good argument for some form of SPF on exposed skin in winter, because UVA damage is cumulative. UVB, the burning, vitamin-D-producing part of the spectrum, drops right down in a UK-type winter, but UVA remains despite the change of seasons, and penetrates into the skin. It chips away quietly at the scaffolding of the skin, year after year, changing pigment, fraying elastic fibres, and creating the background on which skin cancers later arise. That means the apparently trivial exposures – the walk to school, standing on the touchline, light streaming through the office or car window – do add up over decades. This is particularly true for the face, neck, and backs of the hands, which are uncovered even when the rest of you is in a coat and scarf. Building a broad-spectrum SPF into your morning routine, even in winter, is one of the simplest ways to reduce that slow, cumulative UVA toll.

———

DO CAR FUMES AFFECT MY SKIN?

If you live or work next to busy roads, then traffic pollution can undoubtedly cause skin damage. Car and lorry exhausts release a mixture of tiny particles (particulate matter), nitrogen oxides, and other pollutants that sit in the air we breathe. These particles are small enough to cling to the skin's surface, lodge around hair follicles, and interact with the skin's own lipids and microbiome. They generate oxidative stress which, over the years, is linked to more pigment spots, sallowness, rough texture, and fine lines, particularly on the sides of the face that see the most daylight and traffic. In people with eczema, acne, or rosacea, polluted air can also nudge the skin into being a little more inflamed and reactive.

—

HOW DO CIGARETTES DAMAGE THE SKIN?

Every cigarette causes tiny blood vessels in the skin to constrict, so less oxygen and fewer nutrients reach the surface. Over the years, the fibroblasts that produce collagen and elastin skin proteins live in a semi-starved state and create poorer-quality proteins. At the same time, tobacco smoke generates a heavy load of free radicals and triggers enzymes (matrix metalloproteinases) that actively break down the collagen you already have, so the scaffolding of the skin is being dismantled faster than it can be rebuilt. This combination shows up as a sallow or greyish tone to the skin, fine criss-cross lines around the mouth and eyes, deeper folds in the cheeks, and a general loss of elasticity and "bounce". Smoking also slows wound-healing, makes you more prone to infections and visible blood vessels, and is linked to a higher risk of conditions such as psoriasis, hidradenitis suppurativa and, of course, skin cancers. None of this happens after a single cigarette, and some damage is reversible once you stop, but if you are looking for one lifestyle change that will do more for your skin than any cream, serum, or procedure, giving up smoking sits very near the top of the list.

—

06

Guarding the barrier

thinking about your skin

This mini-tool is designed to help you observe and reflect on how your skin behaves. It is not a rigid classification system, but a starting point for building a more informed relationship with your skin.

For each of the domains listed below, read through the three questions and note how many apply to you.

Oily skin

- Do you notice visible shine on your face within a few hours of cleansing, even without applying skincare products?
- Are you prone to blackheads, clogged pores, or dandruff in the scalp or eyebrows?
- Do your scalp or the sides of your nose become flaky or irritated, especially during stress or colder months?

Dry skin

- Does your skin feel tight, itchy, or uncomfortable after cleansing, even without exfoliants or active ingredients?
- Do you experience dull appearances to the skin and dry areas around the mouth or eyes?
- Do you find that moisturizer improves your skin comfort significantly, and that you need to reapply it during the day?

Resilient skin

- Can you use active skincare ingredients (such as retinoids, vitamin C, or exfoliating acids) without stinging, redness, or prolonged irritation?
- Does your skin recover quickly after sun exposure, shaving, or minor abrasions?
- Do you rarely experience intolerance or reactions, even with new or fragranced products?

Sensitive skin

- Does your skin sting, burn, or flush easily when using new skincare products – even those labelled as suitable for sensitive skin?
- Do environmental factors such as wind, heat, or cold trigger visible redness or discomfort?
- Have you been diagnosed with rosacea, eczema, or another condition that affects skin barrier integrity or inflammatory response?

Pigmented skin

- When you get a spot, scratch, or inflamed area, does it tend to darken and remain pigmented for weeks or months?

-)-)-)-
·:·:·:·

- Have you developed areas of uneven pigmentation (such as melasma or sun spots), particularly on your cheeks, upper lip, or forehead after sun exposure?
- Do you tan easily and rarely burn, but develop brown marks from even minor skin trauma?

Non-pigmenting skin

- Do pink or red marks remain on the skin after an acne outbreak has passed?
- Do you burn before tanning or rarely develop a significant tan with sun exposure?
- Does your skin tone remain relatively even year-round, with little variation following sun or inflammation?

Aged skin

- Are you noticing visible changes such as fine lines, volume loss, thinning skin, or reduced elasticity – especially around the eyes, mouth, or neck?
- Do you have a history of significant sun exposure, use of tanning beds, outdoor work without regular SPF protection, or smoke more than 10 cigarettes a day?
- Did your parents or siblings show signs of ageing (wrinkles, sagging) relatively early in life?

Youthful skin

- Does your skin feel firm, smooth, and plump, with minimal fine lines or loss of volume?
- Do you heal quickly from blemishes or minor injury without lasting marks?
- Have you consistently used sun protection and avoided smoking or intensive tanning throughout your adult life?

Conclusions

- If you answered "yes" to two or more questions in a section, you may be displaying features characteristic of that skin profile.
- It is entirely normal to identify with multiple categories. For example, you may have oily skin that is also pigment-prone and sensitive, or dry skin that is also resilient and youthful.
- Avoid labelling your skin as just "dry" or "oily". Most people's skin exhibits overlapping traits, and these may shift over time due to hormones, stress, skincare habits, or age.

Revisit this mini-tool seasonally or annually. Your answers may change depending on where you live, what treatments you use, or how much time you've spent in the sun. The exercise may guide you – and your dermatologist – towards skincare strategies better matched to your biology.

-)-)-)-

behind the labels

The key to understanding skincare is knowing your own skin and its needs. It is also about questioning the claims made on skincare products, many of which are not founded on solid evidence.

Our skin is remarkably self-sufficient. Yet while it does a good job of keeping us alive, it hasn't necessarily evolved to keep us comfortable or radiant as we age. In today's world, we're asking our skin to cope with a far more hostile and unpredictable environment than it was designed for. It is under far more scrutiny than ever before.

In this context, skincare acts as a form of external support, helping skin do its job more efficiently, recover more quickly, and function more comfortably. In addition, skincare can function as a form of self-expression and emotional comfort.

All is not always what it seems

Understanding skincare means making informed choices based on evidence and experience, not fear or fads or ads! It's also important to acknowledge that not everything sold as skincare is grounded in skin science. Many products are not formulated with active ingredients at effective concentrations or in stable forms, and some ingredients are included more for marketing appeal than function.

Much of what's marketed is often more about branding and consumer psychology than dermatological benefit.

-)-)-)-

READING A SKINCARE LABEL

Skincare products may be labelled with impressive-sounding claims, but they are often unregulated, loosely defined, or poorly standardized. Here's what they actually mean.

dermatologist tested
Often means a dermatologist oversaw a skin tolerance test, typically checking for irritation. It does not guarantee that the product treats skin concerns.

clinically proven
The term "clinically proven" has no set standard. It can mean anything from a tiny self-assessment to an uncontrolled in-house test, with no requirement for real scientific proof.

non-comedogenic
Means the product is unlikely to clog pores, but in reality there's no universal standard for testing.

hypoallergenic
Implies a lower risk of allergic reaction, yet this isn't legally defined either. Products may still contain common irritants unless carefully formulated.

anti-ageing
This label is allowed on cosmetics, but it mustn't imply the product alters skin structure or function. If it claims to "reverse wrinkles" or "stimulate collagen", it may cross into drug territory, which is tightly regulated. In most cases, these claims rely on before-and-after photographs or perception studies, not robust clinical trials.

natural
Along with "clean", this claim sounds reassuring but lacks legal definitions. These terms are mostly self-declared by brands and can mean different things depending on who's using them.

medical-grade
There is no official regulatory category called "medical-grade" under UK, EU, or FDA cosmetic law. It's a marketing term, not a scientific classification.

fragrance-free
Usually means no added synthetic fragrance, but some products may still contain natural aromatic extracts (such as essential oils) that can cause irritation.

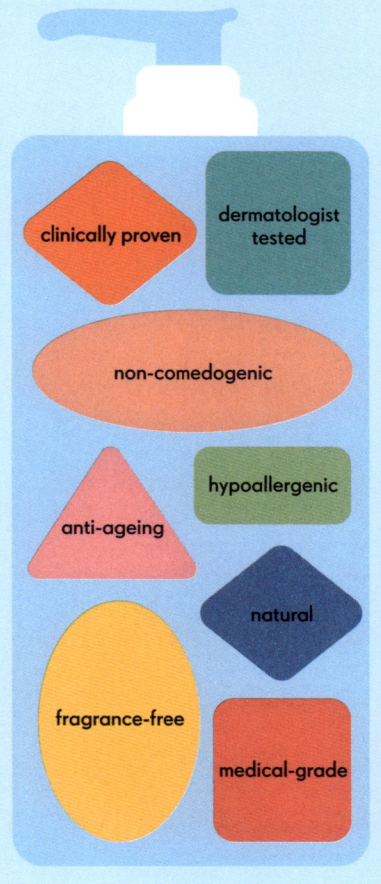

understanding your skin type

We often think of skin type as something we're born with and rarely question whether it can change. Yet what we actually experience on the surface is the result of both biology and biography.

The skin we live in is shaped by more than just our genes. Sun exposure, diet, stress, sleep, pollution, inflammation, and the products we use all leave their mark. Over time, they influence how well the skin repairs itself, holds on to moisture, calms inflammation, and builds the proteins that keep it strong. Our skin type isn't fixed – it's shaped by experience as much as biology.

The four skin systems

Think of your skin as having four main functioning systems: barrier integrity, fibroblast activity, pigmentary reactivity, and sebaceous function. Each of these tells a story, not just about what your skin is like now, but about what it's been through and how it might behave in the future. The best way to understand your skin is by examining these systems to give us more information on how they are working.

BARRIER INTEGRITY

The outermost layer of your skin, the stratum corneum (see pages 18–19), plays a vital role in keeping moisture in and irritants out. A healthy barrier means skin feels calm, comfortable, and resilient. A compromised barrier, on the other hand, often reveals itself through stinging, flushing, or persistent dryness even when there's no visible rash. Some people are born with naturally delicate barriers; others wear theirs down through over-cleansing, harsh products, or repeated inflammation. The difference between a short-term barrier damage (for example, after a laser procedure) and chronically fragile skin (as in eczema) matters clinically. One will bounce back with time, while the other needs careful long-term management.

FIBROBLAST ACTIVITY

Fibroblasts (see page 52) are responsible for the strength, elasticity, and repair capacity of your skin. Fibroblast activity is at its peak in early life but declines with age, particularly in skin exposed to the sun. UV radiation, stress, and inflammation accelerate this decline.

Family history and lifestyle matter here. Did your parents age early or late? Have you had significant sun exposure without protection? These clues help us assess your skin's regenerative potential. Unlike barrier status, which can shift in weeks, dermal ageing is a slower, cumulative process.

-)-)-)-

PIGMENTARY REACTIVITY

Melanocytes, the cells that produce pigment, behave very differently from person to person. Some individuals develop dark marks after every spot, scratch, or patch of inflammation, in a process called post-inflammatory hyperpigmentation (PIH). Others are instead left with redness, especially those with fairer skin types. If you tend to "mark easily" or if acne spots leave a shadow long after they've healed, your pigmentary system is likely highly reactive. Dermatologists look at how long it takes for pigment to fade, how the skin responds to injury, and whether certain areas are more affected than others.

SEBACEOUS FUNCTION

Sebum (see page 15) isn't inherently bad – it helps maintain the skin's barrier and has antimicrobial properties. But too much (or too little) can tip the balance. High sebum output can lead to enlarged pores, blackheads, and acne, in oil-rich areas such as the nose, forehead, and scalp. On the flipside, very low sebum production can result in tightness, flakiness, or even conditions like perioral dermatitis.

Why it matters

Let's take a look at the suggested skincare regimes of two interacting systems.

• Someone with a **fragile barrier**, **pigment-prone skin**, and **high sebum output** may need gentle, non-comedogenic products with pigment-modulating ingredients, barrier support, and oil control.

• Another person with a **robust barrier**, **minimal pigmentary response**, and **low fibroblast activity** might benefit more from collagen stimulation and dermal treatments.

This is why skincare is never one-size-fits-all. A product that works wonders for your friend may leave your own skin red, oily, or inflamed.

Another advantage to this is understanding your skin's tendencies so you can intervene *before* problems arise. For example, if you know you're prone to hyperpigmentation, you can prioritize anti-inflammatory skincare and sun protection. If your fibroblasts are slowing down, you might explore retinoids or energy-based treatments to boost collagen *before* sagging sets in.

-)-)-)-

what is the skin barrier?

Your skin barrier is a highly evolved, multi-layered protective shield — it's a living, layered system that quietly works to keep the outside world out and the inside world in. It plays a central role in how your skin feels, functions, and ages, shaping everything from resilience to radiance.

It is important that you care for your skin barrier in a manner that allows it to function optimally. Clinically, we see the consequences of barrier disruption in virtually every inflammatory skin disorder: atopic dermatitis, contact dermatitis, psoriasis, rosacea, and even acne all involve some level of barrier impairment. Restoration of the barrier is therefore central to treatment.

How do you look after your skin barrier?

Using what we know from clinical and molecular research, the following five steps will help you to maintain a healthy skin barrier.

STEP 1: MINIMIZE DISRUPTION DURING CLEANSING

Choose a low-pH, non-foaming cleanser with mild surfactants (see pages 122–23) such as coco glucoside or decyl glucoside. These remove debris and sunscreen without dissolving the essential lipids that hold your skin cells together. Avoid harsher detergents like sodium lauryl sulfate – they disrupt the lipid bilayers and make your skin more vulnerable to transepidermal

water loss. For most, cleansing once daily in the evening is enough. Your morning skin may only need a rinse with lukewarm water.

STEP 2: REPLENISH ESSENTIAL LIPIDS

Your barrier is built from ceramides (around 50 per cent), cholesterol, and free fatty acids (FFA). After cleansing, apply a moisturizer that contains these in a well-balanced molecular ratio - ideally 3:1:1 of ceramides/cholesterol/FFA. Doing this within 60 seconds of cleansing helps lock in moisture.

STEP 3: PREVENT FURTHER DAMAGE THROUGH SUNSCREENS

UV radiation doesn't just cause pigmentation and collagen breakdown - it destabilizes the barrier itself. It alters lipid composition, disrupts keratinocyte differentiation, and generates oxidative stress. Daily sunscreen with SPF30 or higher is essential, ideally with stable UV filters and antioxidants such as vitamin E or niacinamide.

-)-)-)-

STEP 4: MODULATE ACTIVES TO REDUCE CUMULATIVE IRRITANCY

If your barrier feels compromised, this is the time to pause or reduce any non-prescribed actives. This particularly applies to alpha hydroxy acids, which loosen dead skin cells to smoothen and brighten the skin.

STEP 5: TREAT SYMPTOMS AS SIGNALS

Tightness, stinging, or flushing aren't things to just put up with – they are functional signs of barrier stress. Shift to emollients or barrier-repair creams that contain ingredients such as dimethicone, panthenol, or colloidal oatmeal. Reduce your exposure to potential triggers and resist the temptation to add more layers to your skin routine.

STRONG AND WEAK SKIN BARRIERS

A strong skin barrier retains water, blocks pathogens, and supports healthy microbiomes. Weak barriers lose moisture, allow pathogen entry, and disrupt microbial balance, causing irritation.

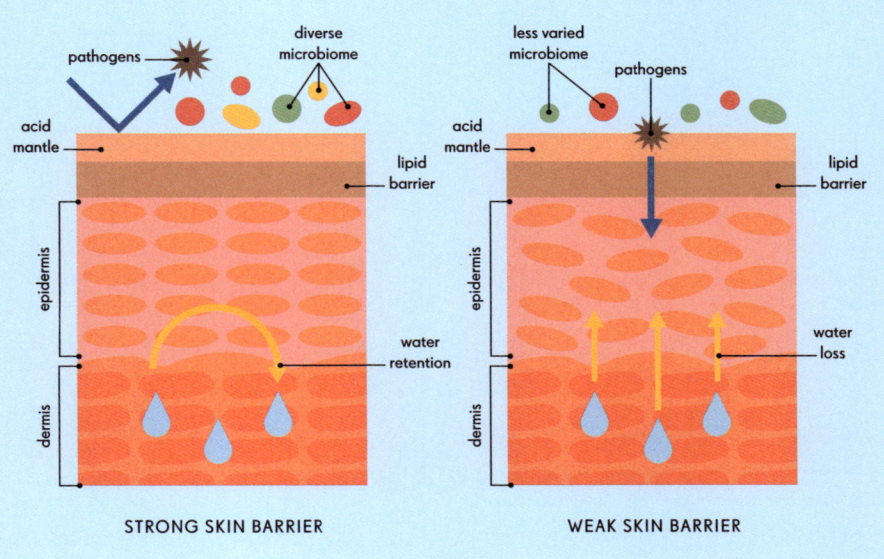

-)-)-)-

building your skincare routine

At its simplest, skincare is built on three essentials: cleansing, moisturizing, and protection. These are your foundation stones — only when they're in place and working well can you start to explore more active, targeted steps.

Skin is translucent, so subtle changes in hydration, smoothness, and surface structure can significantly affect how it reflects light. When your skin is well cared for, it reflects light more evenly, giving off the appearance we associate with "good" skin.

Reflecting on your answers in the Thinking about your skin section (see pages 112–13) can help you make decisions on which products to incorporate in your routine.

Cleansing

Cleansing should be done once in the morning and once in the evening. Over-cleansing disrupts your microbiome and can worsen oil production or sensitivity. Choose a cleanser based on your dominant skin systems (see pages 114–15).

Moisturizing

Your moisturizer's job is to trap water, support the barrier, and deliver helpful ingredients like ceramides, cholesterol, and fatty acids. Usually only very oily skin types can get away without using any moisturizer.

CLEANSING

SKIN TENDENCY	RECOMMENDED CLEANSER
sensitive or barrier-impaired	cream cleanser
oily or acne-prone	gel or foaming cleanser with salicylic acid or an oil cleanser
pigmentation-prone or melasma	gentle balm or milk cleanser
removal of heavy SPF or make-up	oil cleanser

MOISTURIZING

SKIN TENDENCY	RECOMMENDED MOISTURIZER
dry or mature	ceramide-rich, thick cream
oily	gel cream or fluid, non-comedogenic
pigmentation-prone	barrier-repair moisturizer

-)-)-)-

Protecting with SPF

Nothing matters more for long-term skin health than consistent sun protection. Read the feature on this carefully (see pages 126-27) and look for formulations to best support your skin.

Active ingredients

Once your cleanser, moisturizer, and SPF are working well, and your skin has had at least 4-6 weeks to adjust, you may be ready to introduce active ingredients (actives). These are the targeted treatments that help manage chronic concerns such as acne, pigmentation, ageing, and inflammation. Note that adding actives too early or all at once is the most common reason people abandon their routine.

Choose your first target

Look back at the skin systems section of Understanding your skin type (see pages 116-17) and determine your skin's most persistent issue. Refer to the table below to see how to match active ingredients to your specific concerns.

CONCERN	ACTIVE INGREDIENT	WHEN TO USE	NOTES
hyperpigmentation	azelaic acid, niacinamide, tranexamic acid, vitamin C	morning	azelaic acid can also be used in the evening; vitamin C may sting at first
redness	azelaic acid, niacinamide, allantoin	morning or night	start with azelaic acid 10%, move to 15–20% if tolerated
blocked pores	salicylic acid, azelaic acid, adapalene	evening	adapalene 0.1% is the only over-the-counter retinoid and ideal for acne
uneven texture/ dullness	lactic acid, PHA, low-strength retinol or retinal	evening	start with 1–2 times per week to avoid irritation
fine lines, sun damage	tretinoin, retinal, peptides	evening	prescription tretinoin gives the strongest results, but start low and slow
dehydration and barrier weakness	hyaluronic acid, panthenol, ceramides	morning and/ or night	these are supportive and can be layered around actives

-)-)-)-

cleansing

Cleansing doesn't have to follow one script, and for some skin types or lifestyles it is very personal. Let's take a look at the best ways to cleanse your skin and what to avoid.

We cleanse every single day, yet it is something most of us were never actually taught how to do properly. It's also become overly complicated. The sheer variety of ways we can cleanse our face can feel confusing – foams, oils, balms, waters, bars, brushes, cloths – and the simplicity of the old-fashioned "soap and water" routine almost starts to feel appealing again.

That method had its merits, and for some skin types it worked well. But for most, it's probably a little too basic, and can strip the skin if used daily. On the other side, we now have cleansers packed with all kinds of ingredients – acids, actives, plant extracts – some of which are more about marketing than meaningful skin benefits.

The truth is, a cleanser only needs to do one thing well: remove impurities without damaging the skin barrier. Everything else – treating acne, pigmentation, or ageing – is a separate process. Cleansing is functional, and when done correctly, it lays the foundation for everything else. Keep it simple, effective, and kind to your skin.

What is a cleanser?

A cleanser helps remove dirt, oil, debris, and product residue from the surface of the skin. It does this by incorporating ingredients known as surfactants. These tiny molecules bind to grease and impurities with their hydrophobic (water-repelling) tails and rinse away with water thanks to their hydrophilic (water-attracting) heads. Yet surfactants can also remove the natural fats and moisturizing factors in our skin.

Types of cleanser

Classic soaps are made by combining fats or oils with a strong alkali – usually sodium hydroxide – in a process called saponification. This creates a bar of soap that's good at breaking down grease and grime. But it has a high pH, often around 9 or 10, which is far more alkaline than your skin's natural pH (around 5.5). That alkalinity can disrupt the barrier, leading to dryness and irritation.

Syndet cleansers (short for synthetic detergents) are made using gentle cleansing agents rather than traditional soap. Unlike classic soap, which is alkaline, syndets are formulated to match the skin's natural pH, making them less disruptive to the skin barrier. You'll find them in many forms: bars, gels, creams, and lotions.

Foaming cleansers create bubbles that are satisfying, but many traditional foaming cleansers relied on sodium lauryl sulfate (SLS). It's highly effective, but doesn't discriminate

-)-)-)-

between make-up, dirt, and the lipids your skin actually needs, and can strip away the skin's protective oils. Newer formulas use gentler surfactants (such as coco glucoside or sodium cocoyl isethionate) that are much kinder to the skin.

Choosing a cleanser

Cleansers come in various formats, which follow some general rules:

Gel cleansers often suit oily or acne-prone skin, though some can be overly astringent.

Cream and lotion cleansers are richer, with added emollients that make them better for dry or reactive skin.

Oil-based cleansers and balms are especially good at breaking down sunscreen and make-up, and can do so without compromising the skin barrier (unlike many water-based cleansers).

Micellar waters are convenient and gentle, but tend to underperform when used alone, particularly if you wear SPF or heavier skincare.

Choose a cleanser that is:
- **pH-balanced** (around 5.5) to match the skin's natural acid mantle
- **Fragrance-free** or low in fragrance, especially for sensitive skin
- **Free from harsh alcohols** and aggressive surfactants such as sophorolipids (SL)

A note on cleansing wipes. There's no denying that they are convenient, but they're not without drawbacks – most wipes rely on ingredients that can be irritating if left on the skin without rinsing.

HOW SURFACTANTS WORK

Hydrophobic tails surround and embed into oils and dirt, and hydrophilic heads then pull the now-suspended molecule into the water to be washed away.

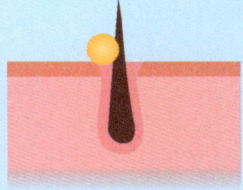

BEFORE
a molecule containing excess oil (sebum) from the hair follicle and dirt attaches to the skin and hair

APPLICATION
when a cleanser is applied to the skin, the tails of the surfactant stick to the molecule

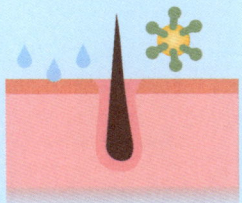

RINSE
when the cleanser is rinsed off, the water-loving heads of the surfactants enable the molecule to be removed with the water

moisturizing

Moisturizers were developed to put lipids back on the skin surface after cleansing. They have now evolved into products to counteract dry skin and deliver cosmetically active ingredients.

Moisturizers are formulated around a trio of ingredients – occlusives, humectants, and emollients – that each have specific tasks (see table below). In practice, most moisturizers contain a mixture of the three ingredients.

Modern moisturizers do much more than just hydrate. Many now deliver active ingredients such as antioxidants, ceramides, peptides, or even sunscreens to support the skin.

The many faces of moisturizers

Moisturizers come in several different forms. Each has its advantages, depending on what your skin needs and where on your body you are applying it.

- **Lotions** are light and often water-based – ideal for normal or oily skin.
- **Creams** are thicker and more hydrating, so better for dry skin or colder climates.
- **Ointments** are grease-based and excellent for severely dry or compromised skin.
- **Gels** are often oil-free and suitable for acne-prone or irritated skin.
- **Serums** deliver active ingredients but don't moisturize much on their own.
- **Balms and sticks** are highly occlusive and great for lips, cuticles, or flaky spots.

THE THREE KEY INGREDIENTS OF MOISTURIZERS

TYPE	WHAT THEY DO	COMMON INGREDIENTS
occlusives	form a barrier over the skin surface to reduce water loss	petrolatum (like Vaseline), lanolin, mineral oil, beeswax, dimethicone, coconut or jojoba oil
humectants	draw water towards them to plump and hydrate skin	glycerine, urea, hyaluronic acid, sorbitol, propylene glycol, honey
emollients	repair barrier "gaps" and soften the stratum corneum	fatty acids (like stearic acid), fatty alcohols (like cetyl or cetearyl alcohol), esters (like isopropyl palmitate), plant oils

-)-)-)-

Rebuilding the skin barrier

When your skin barrier is damaged, a good moisturizer can help to rebuild it. To support barrier repair, look out for moisturizers that contain:

- **Ceramides** – the most abundant lipid in healthy skin.
- **Cholesterol** – another essential lipid that helps maintain flexibility.
- **Fatty acids** – such as linoleic and stearic acid.
- **NMF (natural moisturizing factor)** – a mix of water-binding molecules such as urea, lactic acid, and amino acids.

These ingredients soothe dryness, but also signal to the skin that it can slow down emergency repair responses, helping to restore calm.

Choosing the right moisturizer

The best formulas are tailored to different skin types and environments. For example, If you have oily skin, you might want a lightweight lotion with more humectants and fewer occlusives. If your skin is dry or eczema-prone, you will likely benefit from something richer, containing a petrolatum-based occlusive. If you live in a humid climate, humectants can draw moisture from the air; but in dry environments, they might worsen dryness unless paired with an occlusive.

There's no one-size-fits-all. The right moisturizer for you depends on your skin type, what else you're using (such as retinoids or exfoliants), the climate you're in, and your personal preferences.

- **For dry or mature skin**: Look for thicker creams or balms with high lipid content, ceramides, and cholesterol. Products with ingredients that support ceramide synthesis are especially beneficial.

- **For oily or acne-prone skin**: Gel-creams or lightweight lotions that combine humectants with a small amount of non-comedogenic emollients (like squalane or dimethicone) are ideal. Avoid rich occlusives such as coconut oil or lanolin, which may exacerbate breakouts.

- **For sensitive or reactive skin**: Simpler is better. Opt for fragrance-free, essential oil-free formulations with proven barrier-supporting ingredients like glycerine, shea butter, and ceramides. Patch testing is often helpful.

- **For barrier repair**: Look for formulations that mimic the skin's natural lipids – ceramides, free fatty acids, and cholesterol in a 3:1:1 molecular percentage ratio have been shown to be most effective in restoring barrier integrity.

When and how to moisturize

For most people, moisturizing twice daily – morning and night – is sufficient. Always apply the moisturizer to damp skin after cleansing. This helps trap water in the skin and aids absorption.

In the morning, your moisturizer should be followed by sunscreen as the final step. At night, moisturizer seals in any active ingredients – prescription creams, exfoliants, or serums – applied beforehand.

-)-)-)-

sun protection factor

SPF (sun protection factor) is the most familiar measure on sunscreen bottles. It tells you how well a product protects against UVB, the form of ultraviolet light that causes sunburn and contributes to skin cancer.

UV filters

Sunscreens contain molecules called UV filters. There are two main families:

Organic (or "chemical") filters are carbon-based molecules that absorb UV light and convert it into heat. They tend to be light, transparent, and easy to blend into elegant textures. Examples include avobenzone (excellent UVA1 coverage), octinoxate (a strong UVB absorber), and octocrylene (absorbs UVB and helps stabilize avobenzone).

Inorganic (or "mineral") filters include zinc oxide and titanium dioxide. These are particles that absorb UV light but also scatter about 5 per cent of it. If used together, zinc oxide and titanium dioxide can cover both UVB and the entire UVA range.

The best sunscreens usually use a hybrid approach, combining organic and inorganic filters to optimize protection, texture, and stability.

What is the SPF number?

The SPF number reflects how much longer it takes for skin to burn with the sunscreen compared to without. SPF30 means your skin can tolerate 30 times more UVB radiation before it burns. SPF isn't a perfect block: SPF15 filters out around 93 per cent of UVB rays, SPF30 filters 97 per cent, and SPF50 offers about 98 per cent protection. The higher the number, the more protection, but the returns diminish, and proper application is more important than an SPF100 label.

The question of UVA

It's important to understand that SPF measures protection against only one part of the sun's ultraviolet radiation – UVB. These are the shorter wavelengths primarily responsible for sunburn and direct DNA damage. Yet the other half of the ultraviolet spectrum, UVA, is just as significant. UVA rays penetrate more deeply into the skin and are closely linked to premature ageing, pigmentation, and the breakdown of collagen over time.

If you want meaningful UVA protection, look for sunscreens with additional UVA-specific ratings – like a high "PA" rating – short for "Protection Grade of UVA". This is based on a persistent pigment darkening (PPD) rating system that uses plus signs for simplicity:

- PA+ = basic UVA protection
- PA++ = moderate
- PA+++ = strong
- PA++++ = very high

-)-)-)-

How much SPF should I apply?

Sunscreens work by forming a continuous, even film on the surface of the skin – a protective mesh of UV filters. That film only functions if it's applied in the right amount. The industry standard for testing is 2 mg (0.00007 oz) per square centimetre of skin. That's far more than most people usually apply. If you're using less, you're getting a fraction of the labelled SPF and also missing out on crucial UVA coverage.

For the face, a helpful visual rule is two full fingers' – index and middle finger – worth of sunscreen from base to tip, squeezed out and applied evenly. This closely approximates the tested 2 mg/cm² dose. Because sunscreen films can be a bit like a rack of snooker balls – interlocking, but with gaps between them – a double application can help. Apply one layer, let it settle, and then top up with a second sweep.

When should I apply SPF?

There's often confusion about when sunscreen is necessary. If your aim is to reduce the risk of skin cancer, the key exposure to avoid is UVB – the shorter-wave radiation that's strongest between March and October, typically peaking from around 10:30 am to 3:30 pm. This is when your skin is most vulnerable to direct DNA damage. Yet if your goal is to protect your skin from ageing, pigmentation, or other signs of cumulative damage, then the answer shifts: daily sunscreen on the face becomes essential all year.

Your phone's weather app offers a helpful tool in this decision-making – the UV index.

It provides an estimate of ultraviolet radiation at your location, on a scale from 0 to 11+. While it doesn't distinguish between UVA and UVB, a reading of 3 or above signals that your skin may incur damage if unprotected, even on cloudy days. It's one of the most practical ways to gauge when your skin needs that extra layer of defence.

SPF APPLICATION CHECKLIST

01. Don't get hung up on using the highest SPF. Aim for SPF30 or higher, look for "broad spectrum", and ideally opt for a product with a known UVA rating (like PA++++).

02. Apply on a daily basis to the face and neck; only apply to the body if it is being exposed.

03. For the face and neck, apply the two-finger rule and pat the SPF into the skin.

04. Applying the SPF in two layers is helpful.

In winter and when inside, apply SPF just once a day, but reapply in the summer.

• ADVANCED FORMULATIONS •

Manufacturers use newer techniques like encapsulation, where UV filters are wrapped in microscopic shells to prevent them from interacting with each other to boost stability. Antioxidants are also being added for extra protection to mop up free radicals and reinforce protection.

-)-)-)-

ingredients with evidence

In a world full of promises in skincare, how do you know what to trust? There's a gap between what's marketed and what's meaningful.

In clinic, I'm often asked, "But does it actually work?" It's a fair question. Here are a few ingredients that live up to the hype and are worth knowing about.

Prescribed retinoids

Retinoids are compounds derived from vitamin A that regulate skin cell growth. They boost collagen production, smooth fine lines, unclog pores, and accelerate cell turnover, making them effective for treating acne, hyperpigmentation, and signs of ageing while improving overall skin texture.

Some retinoids are available only by prescription, as they are licensed medicines. That's reassuring: each has been robustly tested, with predictable and reliable outcomes. The most widely used is tretinoin, though other well-established options include adapalene and tazarotene.

Tretinoin, also known as all-trans retinoic acid, is the fully active form of vitamin A. Unlike other retinoids that need to be converted by enzymes in the skin before they can take effect, tretinoin is ready to act immediately. It binds directly to nuclear receptors in skin cells, altering gene expression and changing the way the skin behaves at a cellular level.

In practical terms, this means an increase in cell turnover, a thickening of the epidermis, a reduction in clogged pores, and stimulation of collagen production in the dermis. It also helps to fade pigmentation by speeding up the removal of melanin-containing cells. These effects have been consistently demonstrated in clinical trials and histological studies over more than four decades.

Tretinoin and adapalene are licensed in many countries for the treatment of acne, and tretinoin is also approved for sun damage. In acne, it reduces comedones, dampens inflammation, and helps to prevent future breakouts by normalizing the way skin cells shed into the follicle. In sun-damaged skin, it has been shown to reverse key signs of ageing in human skin. These are not just cosmetic claims but effective treatment for medical conditions.

Over-the-counter retinoids

Most non-prescription retinoid products rely on retinol or retinaldehyde (retinal) – precursor molecules that must be enzymatically converted in the skin to retinoic acid, the active form. This extra metabolic step makes them less potent and their effects tend to appear more gradually.

-)-)-)-

Stability is the key issue with over-the-counter retinoids. Retinol and retinal are fragile molecules – they degrade quickly when exposed to light or oxygen, meaning the ingredient can oxidize in the bottle before it reaches your skin.

Granactive retinoids, such as hydroxypinacolone retinoate, are promoted as low-irritation alternatives, but evidence for meaningful clinical benefit remains limited. Retinyl palmitate and other retinoid esters are even milder, requiring several metabolic conversions before becoming active, which makes their results slow and unpredictable.

Bakuchiol, a plant-derived compound, can influence similar pathways without binding retinoid receptors, but like these gentler derivatives, its effects remain modest compared with prescription retinoids.

Azelaic acid

Azelaic acid is a dicarboxylic acid found naturally in grains like barley and wheat. It inhibits tyrosinase, the enzyme involved in melanin production, making it helpful for pigmentation, particularly in melasma or post-inflammatory marks. It also has antibacterial and anti-inflammatory properties, which makes it remarkably versatile and beneficial in acne and rosacea. In concentrations of 10–20 per cent, it's used to treat acne, rosacea, and hyperpigmentation.

Look for at least 10 per cent concentrations in over-the-counter products or 15–20 per cent in prescription options. Apply once daily to start, then build to twice daily if tolerated.

Niacinamide

Niacinamide, also known as nicotinamide, is the active, amide form of vitamin B3. One of the most reliable multitaskers in dermatology, it has been shown to improve the skin barrier, reduce transepidermal water loss, regulate sebum production, and suppress inflammation. It also helps to fade pigmentation by inhibiting the transfer of melanosomes from melanocytes to keratinocytes, making it a useful ingredient in the treatment of melasma and post-inflammatory marks.

Topical formulations typically contain 2–10 per cent, with 5 per cent being a well-tolerated and effective concentration for most skin concerns. Higher strengths may be useful in oilier skin types but can occasionally cause transient irritation in sensitive individuals.

Salicylic acid (BHA)

Salicylic acid is a beta hydroxy acid (BHA) best known for its ability to exfoliate inside the pores – a unique trait that sets it apart from other acids. It's oil-soluble, which means it can penetrate into the follicle lining, loosening the buildup of dead skin cells and sebum that contribute to comedones and inflammatory acne.

A 2 per cent concentration has been shown to reduce both inflammatory and non-inflammatory acne lesions. Stronger concentrations (of between 10–30 per cent) are typically used in chemical peels under medical supervision for more intensive exfoliation or pigment modulation. For at-home use, 0.5–2 per cent is effective and

-)-)-)-
: : : :

generally well tolerated, though it should be introduced gradually in sensitive or barrier-impaired skin. I often recommend it in cleanser form for oily or acne-prone skin types where it provides contact without lingering too long, and in leave-on formulations for more targeted effects.

I often recommend starting with wash-off formulations to introduce BPO, especially for sensitive or reactive skin. While it can cause dryness, its speed of action and long track record make it one of the few acne treatments that can reduce visible inflammation within days.

Benzoyl peroxide (BPO)

Benzoyl peroxide is one of the most effective treatments for acne and for antimicrobial benefit in the armpits to reduce body odour. Rather than modulating oil production or inflammation alone, it releases free radicals that destroy the bacteria involved in inflammatory acne. Unlike antibiotics, it does not lead to resistance, which is why it remains a cornerstone of acne management, even as prescribing habits evolve.

Sulphur

Sulphur has been used for centuries to treat skin conditions. It helps to dissolve the bonds between dead skin cells, encouraging gentle exfoliation and the clearing of clogged pores. It also inhibits the growth of *Cutibacterium* acnes and *Malassezia* species, both implicated in inflammatory skin conditions. Use it in wash-off formulations – like masks or cleansers – as a gentle but effective way to reduce oiliness and calm inflammation.

SALICYLIC ACID AT WORK

When salicylic acid penetrates pores, it works to exfoliate dead skin cells, dissolve excess oil, and reduce inflammation. It prevents clogging, promoting clearer, healthier skin.

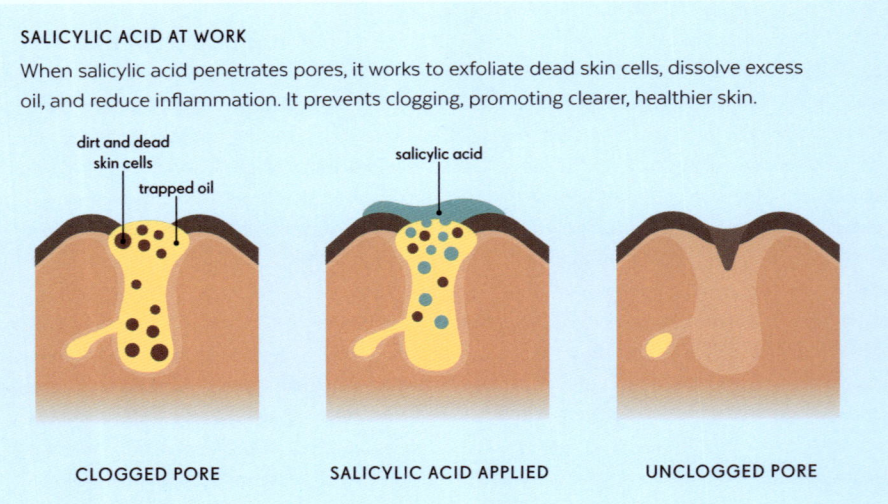

dirt and dead skin cells
trapped oil
salicylic acid

CLOGGED PORE   SALICYLIC ACID APPLIED   UNCLOGGED PORE

antioxidants

The skin is constantly battling against internal and external threats, but antioxidants can help protect cells by preventing some of that damage. Let's take a look at how these substances work.

Sunlight, pollution, cigarette smoke, infrared heat, and even normal cellular metabolism all generate reactive oxygen species (ROS) – unstable molecules that react aggressively with DNA, lipids, and proteins in our skin cells. This is where antioxidants come in, molecules that can intercept and neutralize free radicals (see page 20) before they damage cellular components. The body makes its own antioxidants, and we also obtain many through our diet, but topical application can help reinforce the skin's frontline, especially in the face of modern environmental stress. Importantly, antioxidants do not block UV rays in the way sunscreens do. Instead, they help to limit the damage that UV, pollution, and other exposures cause once they've already interacted with the skin.

Studies have shown that using a well-formulated antioxidant serum can reduce the degree of UV-induced redness (sunburn), lower the activity of enzymes that degrade collagen (such as matrix metalloproteinases), and minimize markers of oxidative DNA damage.

Types of antioxidant

Critically, different antioxidants work at different stages of the oxidative cascade. Vitamin C, for example, scavenges free radicals in the fluid surrounding cells, while vitamin E protects the lipid membranes. Enzymes such as superoxide dismutase neutralize one type of radical, while others like catalase and glutathione peroxidase handle downstream byproducts. This is why a cocktail of antioxidants – rather than a single ingredient at high concentration – is often more effective. They function as a team, each targeting a specific type of damage or step in the process. Just as no single sunscreen protects against every wavelength of light, no single antioxidant neutralizes every type of free radical.

Here are some of the most robust antioxidant options in modern skincare, based on both clinical data and biochemical plausibility.

L-ASCORBIC ACID (VITAMIN C)

Among the most studied and celebrated antioxidants in dermatology, L-ascorbic acid is the active, water-soluble form of vitamin C. It plays multiple critical roles in skin health: it neutralizes reactive oxygen species, helps regenerate oxidized vitamin E, and is essential for the synthesis of collagen. In addition to its antioxidant function, L-ascorbic acid also inhibits tyrosinase, the enzyme involved in melanin production, which makes it useful in

-)-)-)-

managing pigmentation concerns. When applied topically at concentrations between 10 and 20 per cent, L-ascorbic acid has been shown in clinical studies to reduce the visible effects of sun exposure, such as erythema (sunburn), fine lines, and uneven tone.

TOCOPHEROL (VITAMIN E)

Tocopherol is the most biologically active form of vitamin E. It is a fat-soluble antioxidant that integrates into the lipid-rich membranes of our skin cells. There, it acts as a frontline defender against lipid peroxidation – a damaging chain reaction set off by ultraviolet light and environmental stress.

Its true strength, however, emerges in partnership. For example, as vitamin C neutralizes reactive oxygen species in the watery parts of the cell, tocopherol stabilizes the cell membranes. And when either becomes oxidized in the process, the other helps regenerate it. This mutual reinforcement makes them significantly more effective together than alone.

FERULIC ACID

A plant-derived antioxidant, ferulic acid stabilizes both vitamin C and E when combined in formulations. It absorbs UV and visible light and further boosts antioxidant efficacy.

COENZYME Q10 (UBIQUINONE)

This lipid-soluble antioxidant is naturally found in skin but declines with age. Topical application (typically at 0.3–1 per cent) has been shown to reduce wrinkle depth and oxidative stress in photo-aged skin. It's particularly useful in barrier-compromised or dry skin formulations due to its lipophilic (fat-loving) nature.

POLYPHENOLS

Plant-derived antioxidants such as epigallocatechin gallate (EGCG) from green tea, resveratrol from grapes, and polyphenols found in coffeeberry offer a rich and diverse source of free radical defence. These compounds not only have potent antioxidant activity but also bring anti-inflammatory benefits, which can help to calm and protect the skin under stress. While they are not as well studied as vitamins C or E, they offer a promising and increasingly sophisticated layer of protection in modern skincare.

How to use antioxidants

Antioxidants should be thought of as complementary protection. They don't block UV rays – you'll need an SPF to block incoming radiation, and then the antioxidants will reduce the internal fallout from any ROS that slip through. Apply them in the morning, usually directly onto clean, dry skin, under your moisturizer and SPF.

-)-)-)-

peptides

Peptides undertake various beneficial tasks within your skin.
They are an active ingredient in some skincare products,
though their effectiveness is still being evaluated and
needs to be treated with caution.

Peptides are tiny chains of amino acids – usually just 3 to 30 links long – and they work a bit like text messages between cells. Depending on the message they carry, peptides can tell your skin to do a number of things: make more collagen, stop breaking it down, calm inflammation, ferry important minerals into the skin, or relax the micro-muscles that contribute to expression lines. Let's unpack what this all actually means.

Peptides have an appealing scientific rationale, but most evidence is from small or manufacturer-led studies rather than large independent trials. So while early results are promising, their real-world benefits remain unproven and should be viewed with cautious optimism.

Signalling peptides

Signalling peptides are designed to stimulate fibroblasts (see page 52), the cells that build your skin's structural scaffolding. When your body naturally breaks down collagen, tiny fragments circulate and act as a distress signal. Signalling peptides mimic these fragments and trick the skin into thinking more collagen is needed. A well-known example is palmitoyl pentapeptide-4 (you'll often see it under the brand name Matrixyl).

Carrier peptides

Carrier peptides deliver essential minerals into the skin. Copper is a key player in wound-healing, and copper peptides such as GHK-Cu play a role in healing injured or aged skin by activating genes involved in tissue regeneration and calming down the enzymes that chew through collagen.

Enzyme-inhibiting peptides

These peptides work by disabling the enzymes that break down collagen and elastin. Clinical evidence here is less robust than with signalling peptides, but promising.

Neurotransmitter-inhibiting peptides

These "Botox-like" peptides are designed to reduce muscle contractions that lead to wrinkles, but are not as effective as botulinum toxins (see pages 152–53). Argireline is the best-known type.

If you want to try peptides, choose reputable brands that share ingredient concentrations and use proven delivery systems. Apply them after cleansing on slightly damp skin.

-)-)-)-

HOW DO I READ THE INGREDIENT LIST ON A PRODUCT?

Ingredient lists follow a very simple rule in that everything is listed in descending order of concentration until you reach one per cent, after which the order can be mixed. Names are written in INCI (International Nomenclature of Cosmetic Ingredients), which is designed for regulatory clarity, not consumer friendliness. Fragrance may be listed as "parfum" on a label, yet it often masks a very complex mixture of chemicals that can be irritating for some people. Percentage of the ingredient is rarely disclosed – too low (in most cases!) and the ingredient is decorative, too high and it risks irritation. Even when the percentage is perfect on paper, the formula itself determines whether an ingredient penetrates, remains stable, and actually works. Two serums with the same ingredient list can perform very differently depending on pH, encapsulation, solvent choice, and manufacturing.

———

IF I HAVE AN ALLERGY TO AN INGREDIENT, HOW DO I KNOW WHAT TO DO TO AVOID IT?

If you have a confirmed allergy after patch testing, it becomes essential to check the ingredient list every time you buy a new product. You will be given the chemical name of the allergen, but what makes things confusing is that one ingredient can appear under several different names. I often suggest patients use the website incidecoder.com. There you can enter the exact product and it will show you the current ingredient list, the other names that ingredient might go by, and any recent formulation changes. Companies can update preservatives or tweak formulas without making a major announcement, so if you suddenly react to a product that used to be fine, it is worth checking whether the formulation has changed. It is also helpful to know that some substances cross-react because they share a similar molecular structure – for example, PPD in hair dye and hydroquinone in some skin-lightening products. Other ingredients only become a problem when they are exposed to sunlight.

———

WHAT IS THE BEST WAY TO START USING A RETINOID?

Usually, the first step is to work out which, if any, retinoid is appropriate for your skin. For people with acne, rosacea, melasma, hyperpigmentation, or more advanced signs of ageing such as deeper wrinkles, a conversation about prescription retinoids is important, because that is where the strongest evidence lies for treating these conditions effectively. You will find few dermatologists who don't have retinoids near the top of their prescribing list. For adults whose main concern is fine lines rather than a diagnosed skin condition, over-the-counter retinal or retinol can be a place to start. Concentration matters – the dose your skin actually receives will dictate how effective the treatment is. The treatment is applied once in the evening in a thin layer across the face and avoiding the mouth, nose, and corners of the eyes. Irritation can occur, and though it isn't always a sign to stop, just take a bit of care – avoiding that area for a few days may be helpful.

———

ARE STRONGER DOSES OF RETINOIDS BETTER?

Prescription retinoids are made in fixed strengths for specific conditions. In cosmetic products, the rules are looser, and the percentage on the label can sometimes be misleading. Retinoids can be stronger in two ways: concentration and the molecule itself. For the latter, retinoic acid (tretinoin) is intrinsically more potent than retinaldehyde, which is more potent than retinol, which in turn is more potent than retinyl esters. Non-prescription forms (any type other than tretinoin) need to be converted to retinoic acid, which is done by enzymes in the skin that convert a tiny amount and move the rest to storage. But there is only a finite effect of the enzymes, so once they are working flat-out you aren't getting any more conversion to the active form – it isn't doing anything more for your skin. So, doses of retinol around 0.1% seem to be the sweet spot; anything stronger is not getting converted to the active molecule and ends up causing irritation. With prescription retinoic acid, the conversion steps are bypassed and the molecule binds directly to the receptor. There is no "ceiling dose" here – the optimum strength of tretinoin depends on the patient and the condition being treated.

———

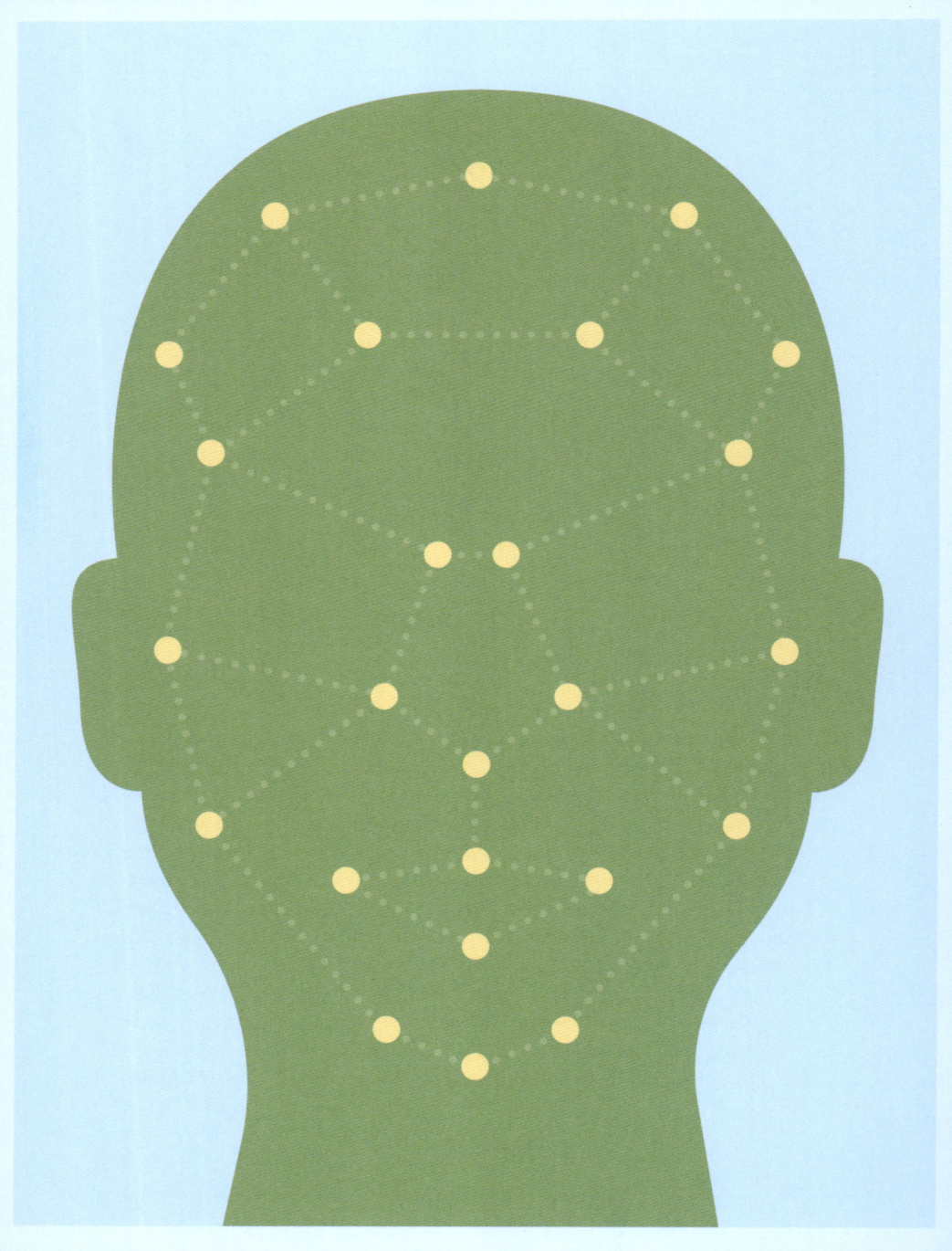

07

The clinic

first impressions

There are occasions in life that lead you to sit in a bright white room, surrounded by laminated posters of moles and rashes waiting to see a specialist. You're now in a dermatology clinic, and this chapter is your backstage pass!

People come to dermatologists for many reasons – perhaps your GP has referred you or maybe you've found your own way here after an online search and good intentions.

What to expect

When you first attend the clinic, it's good to be prepared. It can be helpful to write down the questions that have been worrying you, so you don't forget to ask them during the consultation. Bring a list of any regular medications you're taking, along with details of any creams, tablets, or treatments you've already tried for your skin. If your rash or skin concern has changed over time, photos can be very useful.

Dermatologists usually need to examine all of your skin, not just the area you're concerned about, so you may be asked to undress. If you feel more comfortable with a chaperone present, you can absolutely request one. Alongside your dermatologist, you may also meet dermatology trainees, specialist nurses, administrative staff, or, in teaching or research hospitals, researchers involved in clinical studies.

Skin imaging techniques

In dermatology, we're fortunate in that the skin is right there in front of us – its signs and symptoms are visible to the eye. Even so, we often rely on skin imaging tools to see the defining features of certain conditions more clearly, and sometimes even to visualize the cells themselves.

DERMOSCOPY

The most common tool in the dermatologist's kit is the dermatoscope. It's a small, handheld device that looks a little like a cross between a torch and a magnifying glass. It combines magnification with polarized light to give a clearer view of the structures beneath the surface of the skin. With it, we can examine moles, rashes, and lesions in far greater detail than with the naked eye. It is particularly useful in the diagnosis of melanoma, as it allows us to identify specific patterns within a mole that are more in keeping with early melanomas. This helps us avoid unnecessary surgery for harmless lesions, while also ensuring we detect melanomas at the earliest possible stage.

PHOTOGRAPHY

We use high-quality images to document skin lesions, monitor changes over time, and support diagnosis. In patients at higher risk of skin cancer, we may also use total body photography – a full set of images are taken to track new or changing moles. Increasingly, new tools are incorporating AI-assisted mapping and lesion tracking, helping us detect subtle changes earlier.

REFLECTANCE CONFOCAL MICROSCOPY (RCM)

This non-invasive imaging technique allows us to view the skin at cellular resolution, in real time. Often described as a "virtual biopsy", RCM uses a low-power laser to scan just beneath the skin's surface, producing detailed black-and-white images of skin architecture. It's particularly useful in diagnosing certain skin cancers, and in evaluating lesions where a traditional biopsy might be cosmetically sensitive or unnecessary. It doesn't replace all biopsies, but RCM can help avoid them in some cases, guide the best place to biopsy, and monitor treatment responses.

OPTICAL COHERENCE TOMOGRAPHY (OCT)

This is used to create cross-sectional images of the skin, a bit like ultrasound but using light instead of sound. It allows us to see the layers of the skin, helping to assess the composition of lesions without removing tissue. While it doesn't offer the detail of a traditional biopsy or RCM, it provides valuable structural information.

REFLECTANCE CONFOCAL MICROSCOPE

A near-infrared laser focuses light into a layer of the skin; reflections of that layer pass through a pinhole, creating a detailed image.

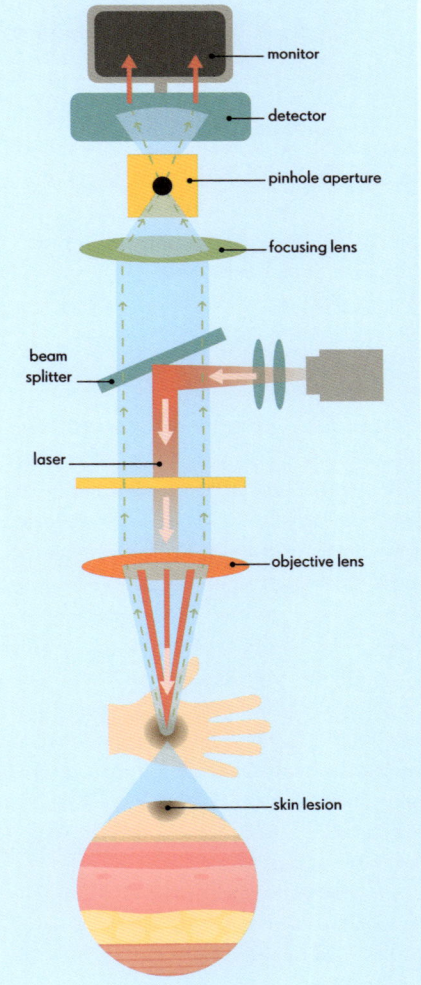

- monitor
- detector
- pinhole aperture
- focusing lens
- beam splitter
- laser
- objective lens
- skin lesion

sampling the skin

One of the great advantages of dermatology is that the organ we work with is right there on the surface and, when needed, we can sample it with relative ease. This process is often called a biopsy, although "sample" is sometimes used as a more approachable term.

We sample the skin for many reasons. It might be to find out whether a rash is caused by inflammation, infection, or malignancy. Sometimes it's to confirm a clinical suspicion, and other times it's the only way to solve a diagnostic puzzle when the surface alone doesn't tell the full story.

Swabs – sampling for infection

When we suspect that a bacterial or viral infection may be involved, the first and simplest test is often a skin swab. A sterile cotton or polyester-tipped swab is gently rubbed over the surface of the lesion or rash. For bacteria, the swab is typically placed into a special transport medium, a small tube containing a jelly that helps preserve any bacteria present on the sample until it reaches the lab. There, it's plated onto culture media and incubated to allow the bacteria to grow. Once colonies are visible, they can be examined under a microscope and tested for antibiotic sensitivities.

Viral swabs work similarly but use a viral transport medium, designed to keep fragile viruses stable during transport. This is especially useful when we're trying to identify conditions such as herpes simplex, varicella zoster, or human papillomavirus. The virus is then identified using PCR (polymerase chain reaction) techniques.

For fungal infections, we gently scrape the affected skin (usually with a blunt scalpel or glass slide) to collect flakes of the outermost layer. These skin scrapings are then examined under the microscope, sometimes after adding potassium hydroxide (KOH), to dissolve skin cells and highlight the fungal structures. The species of fungus can then be identified, and targeted antifungal treatment recommended.

Biopsies – looking beneath the surface

When we need to understand the underlying structure of the skin, we take a biopsy. This involves removing a small piece of skin so it can be processed and examined under the microscope by a dermatopathologist. It allows us to look at the organization of the cells, their shape and size, and where exactly in the skin the problem lies: the epidermis, the dermis, or deeper in the subcutaneous fat.

-)-)-)-

PUNCH BIOPSY

One of the most common methods is a punch biopsy. Here, a small, circular blade, usually 2–8 mm (0.08–0.31 in) in diameter, is used to remove a cylindrical core of skin, a bit like a miniature apple corer. This gives us a full-thickness sample, from the surface then preserved in formalin (a liquid form of formaldehyde) and sent to the laboratory.

OTHER BIOPSY TYPES

Another technique is the shave biopsy, which involves using a scalpel to remove just the top layers of the skin, often used for superficial lesions or when we don't need to sample the deeper dermis. For larger or more complex lesions, an incisional biopsy (removing part of the lesion) or an excisional biopsy (removing the whole lesion) may be performed.

Under the microscope

Once the biopsy reaches the lab, it's embedded in wax, sliced into ultra-thin sections, and stained – most commonly with haematoxylin and eosin (H&E), which highlights different tissue components in shades of pink and purple. The slide is then reviewed under a microscope, allowing us to identify characteristic patterns of disease. This level of detail is invaluable – it helps us confirm diagnoses and get to the right treatment more quickly.

> **Different conditions leave very specific signatures in the skin – psoriasis, lupus, eczema, and skin cancers all have recognizable microscopic hallmarks.**

identifying allergens

Allergies are one of the most common conditions in modern
life, from dairy and gluten to skincare and household dust.
Yet it's important to differentiate between discomfort,
intolerance, and a true allergic reaction.

Knowing whether you have a real allergy to something or not can spare you unnecessary restrictions or even a medical emergency. That's why getting the right tests is so important.

In dermatology, our focus is often on allergens that affect the skin by evoking either a:

• Type 1 (immediate) hypersensitivity reaction, e.g. house dust mite allergy, which is caused by an immune response to proteins in the bodies and droppings of *Dermatophagoides* species microscopic mites that live in house dust.

• Type 4 (delayed) hypersensitivity reaction, e.g. an allergy to nickel, one of the most common causes of allergic contact dermatitis. People who are sensitized to nickel develop a delayed eczema-like rash at sites where the metal has come into prolonged contact with the skin.

Testing for allergies

The world of allergy testing is vast and, frankly, a little confusing. There are in-clinic tests, home kits, blood tests, elimination diets, and even smartphone apps promising to tell you what's causing your symptoms. But not all of these approaches carry the same level of scientific rigour. Some are based on decades of evidence, while others can produce misleading results that create anxiety or lead people to avoid perfectly safe foods, medications, or environments. Here are some of the options available.

SKIN PRICK TESTING

This is most commonly used by allergists to identify immediate allergic reactions, particularly to foods, environmental allergens, or insect stings. It's fast, relatively painless, and can provide results within 15 to 20 minutes.

Here's how it works. A drop of the suspected allergen is placed on the skin (usually the forearm), and then a small lancet is used to gently prick the skin through the drop. If you're allergic, a small red bump, similar to a mosquito bite, will appear within minutes. It's highly sensitive and useful when you need to identify reactions that might lead to more serious systemic symptoms.

However, it's not particularly useful for identifying skin-related allergies like contact dermatitis; for that, patch testing is far more accurate. And, again, a positive result doesn't always mean clinical relevance – it must be interpreted in the context of real-life reactions.

-)-)-)-

PATCH TESTING

Patch testing is the gold standard to identify delayed hypersensitivity reactions such as allergic contact dermatitis. In clinic, small amounts of up to 100 different allergens are applied to your back using specially designed adhesive patches. These are left in place for 48 hours, during which time you must avoid getting the area wet or sweaty. The patches are then removed and your skin is examined, and then again at 72–96 hours to see whether a reaction has developed.

A positive result usually occurs when there is a red, raised, itchy area at the site of one of the particular applied allergens; in an especially strong reaction there can also be blistering. This is how to identify the culprit!

IGE BLOOD TESTS

Blood tests that measure immunoglobulin E (IgE) levels are often used in allergy testing. IgE is an antibody produced by the immune system that plays a key role in allergic reactions. When triggered, IgE prompts the release of histamine and other chemicals that cause the familiar signs of an allergic reaction: itching, redness, swelling, and sometimes more serious symptoms.

The tests can check for a particular allergy. For example, in the case of tree pollen then the specific IgE for tree pollen would be elevated. But sometimes IgE levels are raised across the board, as often happens in eczema. This can make it look as if you're allergic to everything, when in fact your immune system is simply on high alert. For that reason, IgE blood test results should always be interpreted alongside your clinical history.

HOW TO PATCH TEST AT HOME

The AAD recommends the following steps for patch testing a new skincare product:

1. Apply the product to a small patch of skin where you are unlikely to accidentally wash or rub it away. Ideal areas include the inside of the arm or bend of the elbow. Apply the product to a 10p coin-sized patch of skin. You should apply the product as thickly as you would when using it regularly.

2. Leave the product on the patch of skin for as long as it would normally be on the skin. If you are testing a product that you would usually wash off, such as a cleanser, keep the patch on for 5 minutes – or as long as the instructions advise – before washing off.

3. Repeat the patch test twice a day for between 7–10 days. A reaction may not happen immediately, so it is important to continue applying the product for this length of time.

4. If your skin reacts to the product, wash the product off as soon as possible and stop using it. You can use a cool compress or petroleum jelly to relieve the skin if needed.

diagnosing hair concerns

Our scalp and hair are part of our skin system. When problems
arise — such as hair loss, scaling, itching, or inflammation — a
dermatologist may carry out a series of targeted examinations
and tests to identify the underlying cause.

As with all good medicine, diagnosing hair
concerns begins with a detailed clinical history.
The dermatologist will ask when the issue started,
how quickly it's progressed, whether it's come
and gone, and if there are other symptoms
such as itching, tenderness, or flaking. You
may be asked about your general health, recent
illnesses, medications, stress levels, dietary
habits, hormone changes (like pregnancy or
menopause), and any family history of hair loss.

These clues help us narrow down the long
list of potential causes. Using the illuminated lens
of a dermatoscope (trichoscopy), we can see
diagnostic patterns not visible to the naked eye.

What does a
dermatologist look for?

A thorough examination starts with the pattern
of hair loss. For example, general thinning and
receding of the hairline suggests androgenetic
alopecia, whereas discrete round or oval smooth
patches indicate alopecia areata. We also assess
the condition of the scalp itself, looking for
patches with redness, scale, sebum, pustules,
tenderness, or signs of infection. Redness,
particularly perifollicular (around the hair follicle),

often points to inflammation at the follicular level,
common in autoimmune or scarring conditions.

Loss of follicular openings is a key sign of
scarring alopecia - once the follicles are
destroyed, hair regrowth is no longer possible.
Eyebrows, eyelashes, and body hair are reviewed
when indicated. This structured approach helps
distinguish non-scarring conditions from scarring
processes that require urgent intervention.

Specific test types

In some cases, additional tests are used to
confirm a diagnosis or assess the severity
of the problem.

- **Hair pull test**: gently tugging a small bundle of
  hair from the scalp can reveal whether excessive
  shedding is occurring, and from which area.

- **Hair wash test**: this can be done at home,
  collecting shed hairs over a set time to assess
  the quantity and type.

- **Scalp biopsy**: when the diagnosis remains
  uncertain, we typically take a 4 mm (0.16 in)
  punch biopsy (see page 141) from an active

area, ideally containing both affected and unaffected follicles. Two samples are usually taken: one for horizontal (transverse) sectioning to examine hair follicles across the dermis, and one for vertical sectioning to assess the overlying epidermis and dermal interface.

- **Blood tests**: these may include checks for iron deficiency (ferritin), thyroid function, vitamin D, B12, hormone levels (testosterone, DHEAS), or autoimmune markers, depending on the suspected cause.

Matching the pattern to the diagnosis

There are many types of hair loss, but the most common ones we diagnose include:

- **Androgenetic alopecia** (pattern hair loss): this is hormonally and genetically driven. It causes gradual thinning over the top of the scalp in both men and women.

- **Telogen effluvium**: a temporary increase in hair shedding, often after illness, stress, childbirth, or major life events. It usually resolves within 6–9 months.

- **Alopecia areata**: an autoimmune condition where the immune system attacks hair follicles, causing smooth, circular bald patches. It can be self-limiting or progressive.

- **Scarring alopecias**: these are rarer and involve permanent follicle destruction. Early diagnosis and treatment are essential to prevent irreversible hair loss.

- **Traction alopecia**: caused by tight hairstyles over time, this is often seen along the hairline.

- **Tinea capitis**: A fungal infection of the scalp, most common in children, causing patchy hair loss with scale or pustules.

> ### • A TRICHOGRAM •
> Sometimes the problem is not in the follicle, but in the hair shaft itself. A trichogram is when plucked hair is visualized with light microscopy. It is a useful technique to quantify hair cycle abnormalities and diagnose specific causes of hair loss.

-)-)-)-

biologics

Biologic drugs are a newer type of medicine, made
using living cells, rather than from chemicals in a
lab. They are designed for complex conditions
where standard treatments are not working.

In skin disease, biologics have transformed lives.
When I first started in dermatology, people with
severe psoriasis or eczema often spent weeks in
hospital trying light treatments, steroid creams,
or tablets with harsh side effects. Now, almost
all of those same patients can be managed at
home on biologics. Some go from being unable
to sleep, work, or even wear clothes comfortably
because of constant itch and pain, to having
clear skin and a normal life.

These medicines don't cure the condition, but
they control it so well that for many people the
disease no longer controls them.

How do biologics work?

Most traditional drugs used in severe skin
disease – such as methotrexate or ciclosporin –
dampen down the whole immune system.
Biologics are far more precise. They block just
the parts of the immune system that are
overactive in a particular disease, leaving the
rest to do its normal job of fighting infections.

For example:

• In psoriasis, certain messenger proteins called
cytokines (IL-17, IL-23, TNF) drive skin cells to
grow and inflame too quickly. Biologics can
block those exact signals.

• In eczema, different messengers (IL-4 and
IL-13) keep the immune system switched "on",
causing redness and itch. A biologic can target
those two proteins directly.

• In hidradenitis suppurativa (a painful condition
with deep boil-like lumps), blocking TNF helps
calm the inflammation.

Because they act at the root of the problem,
biologics are often very effective and usually
cause fewer whole-body side effects. They're
usually given by injection under the skin every
few weeks, or occasionally by a drip in hospital.

-)-)-)-

The main biologic types

Most dermatology biologics fall into two groups:

- "-mab" drugs (mab is short for monoclonal antibodies): these are lab-made proteins that act like keys fitting into very specific locks on immune cells. Once attached, they block the faulty signal that drives disease. Examples include secukinumab, which blocks IL-17 in psoriasis, and dupilumab, which blocks IL-4 and IL-13 in eczema.

- "-cept" drugs (receptor fusion proteins): instead of a key, think of these as a sponge. They soak up the harmful signals before they can cause damage. For example, etanercept traps an inflammatory protein called TNF (tumour necrosis factor), and is used for psoriasis and psoriatic arthritis.

Before taking biologic drugs…

Biologics aren't the first step in treatment. They are usually offered only when someone has moderate to severe disease that hasn't responded to creams, light therapy, or tablets. Before you can start, your doctor will:

- Check for infections such as TB and hepatitis.

- Do blood tests to make sure your immune system and liver are healthy.

- Talk through the risks (such as a slightly higher chance of infections) and how to monitor them.

Targeting a narrower part of the inflammatory pathway treats the disease more precisely, with fewer whole-body immune side effects.

-)-)-)-

phototherapy

Phototherapy refers to the medical use of ultraviolet radiation to treat skin disease. Its aim is to reduce inflammation, slow down overactive skin cells, and calm the immune response.

Having explained the damage ultraviolet radiation can cause to the skin (see pages 18–19), it may be surprising to learn that we also use it to treat certain conditions. In a carefully controlled medical setting, ultraviolet light can be harnessed therapeutically. It remains one of the most effective, evidence-based treatments for several chronic inflammatory skin diseases – a field known as phototherapy.

Unlike sunlight, which delivers a broad and unpredictable mix of UVA and UVB, phototherapy isolates the wavelengths shown to help. This includes narrowband UVB, the most commonly used form today, and UVA, used for deeper or more resistant conditions.

Varieties of phototherapy

NARROWBAND UVB (NB-UVB)

Narrowband UVB uses a small segment of the UVB spectrum (311–313 nanometres) that has been shown to produce optimal anti-inflammatory effects with fewer side effects. This wavelength suppresses overactive immune cells in the skin, and is useful for conditions such as atopic eczema.

Patients typically attend treatment 2–3 times a week over 6–10 weeks. It is delivered in a walk-in phototherapy cabinet or, in some cases, with handheld or localized devices. Each session lasts only a few minutes, with the dose gradually increased based on skin type and tolerance.

PHOTOCHEMOTHERAPY (PUVA)

PUVA is a more intensive form of phototherapy that combines UVA light with a photosensitizing agent – which makes the skin more sensitive to sunlight – called psoralen. This can be given orally (as tablets taken two hours before treatment) or applied directly to the skin (via a bath or soak). Once in the system, UVA light is delivered; this activates the psoralen, which slows cell division and turns off overactive immune cells in the skin.

Because PUVA penetrates further into the skin, it is often reserved for more severe or treatment-resistant cases, such as advanced psoriasis. It carries a higher long-term risk of skin cancer, especially with high cumulative doses, and is generally avoided in children. Patients receiving PUVA must wear UV-blocking sunglasses for 24 hours after treatment to protect their eyes from psoralen-sensitized UVA exposure.

-)-)-)-

BATH PUVA

In some patients, disease is limited to the palms and soles, and a modified version of PUVA – known as bath or soak PUVA – is used. The hands and/or feet are soaked in a dilute psoralen solution for 15 minutes, then exposed to UVA light in a specialized, localized unit. This allows targeted treatment without exposing the rest of the body to unnecessary UVA.

EXCIMER LASER

The excimer laser is a highly targeted form of phototherapy that delivers the 308 nm wavelength of UVB light to small, defined areas of skin. It is especially useful for localized conditions such as vitiligo, particularly in cosmetically sensitive areas. Because of its precision, excimer therapy allows higher doses to be delivered directly to affected skin while sparing surrounding healthy tissue, reducing side effects and treatment time.

PHOTODYNAMIC THERAPY (PDT)

Photodynamic therapy is a distinct type of light-based treatment used primarily for precancerous lesions, including actinic keratoses, superficial basal cell carcinomas, and occasionally acne. It involves applying a light-sensitizing cream (usually 5-aminolevulinic acid or methyl aminolevulinate) to the skin. After an incubation period (typically 1–3 hours), the area is exposed to a specific wavelength of visible light, activating the photosensitizer. This generates reactive oxygen species that selectively destroy abnormal or sun-damaged cells.

Are there any risks to phototherapy?

Long-term risks include photo-ageing and, in the case of PUVA, an increased risk of skin cancer if very high cumulative doses are given. Narrowband UVB, in contrast, has not been associated with a significantly increased skin cancer risk, even with long-term use, and is considered safe for children and pregnant individuals.

• A WORLD AWAY FROM TANNING SHOPS •

Going to a local sunbed shop is not the same as having medical phototherapy. In clinic, our phototherapy units are carefully regulated, regularly tested, and designed to deliver only a very precise, narrow band of UVB that treats the skin safely. Commercial sunbeds don't offer that level of control or calibration – without proper monitoring, they can expose you to the wrong wavelengths and much higher doses, which can genuinely cause harm.

-)-)-)-

cryotherapy

Cryotherapy is a quick, simple, and effective treatment that involves the application of extreme cold to destroy abnormal or unwanted skin tissue. Let's take a look at what it involves and which conditions it can help.

The goal of cryotherapy is to cause cellular destruction through freezing, followed by a localized inflammatory response that helps clear the treated lesion. This method is particularly effective for superficial lesions and is commonly used to treat:

- Viral warts (including common and plantar warts)
- Actinic keratoses (sun-damaged precancerous lesions)
- Seborrhoeic keratoses
- Molluscum contagiosum
- Some small superficial skin cancers, particularly superficial basal cell carcinoma in low-risk areas

"Freezing" agents

The most common "freezing" agent used in cryotherapy is liquid nitrogen, which reaches temperatures around −196°C (−321°F). Over-the-counter (OTC) cryotherapy products, often marketed for wart or skin tag removal, typically use chemical agents such as dimethyl ether and propane, which cool down to around −50°C (−58°F). It means that OTC products freeze only the uppermost layers of the skin and are generally less effective.

The cryotherapy process

Cryotherapy is a minor procedure usually carried out in a dermatology clinic. It involves freezing the lesion using either a cryospray canister or an applicator dipped in liquid nitrogen.

The area is frozen for between 5 and 30 seconds depending on the type, depth, and location of the lesion. The most common method is the "freeze-thaw-freeze" technique, in which the lesion is frozen, allowed to thaw, and then frozen again to enhance tissue destruction.

After cryotherapy, the treated area is usually darker in colour and may become firm until it naturally "drops off". Occasionally, a blister forms, which scabs and peels over the next one to two weeks. Topical antiseptics may be recommended.

Most patients tolerate the procedure well, although a brief stinging sensation during freezing and for a few hours afterwards is common. In darker skin types, cryotherapy carries a higher risk of post-inflammatory pigmentary change – either lightening or darkening – and, more rarely, scarring. These potential side effects are discussed as part of the initial consultation process between the clinician and the patient.

-)-)-)-

chemical peels

Chemical peels are a method of resurfacing the skin by applying a chemical solution that exfoliates and removes layers of sun-damaged or pigmented tissue. This encourages regeneration, potentially improves scarring, and reduces signs of photo-ageing.

Chemical peels sit at the intersection of cosmetic and medical dermatology. They can be used to refresh dull skin, soften fine lines, reduce pigmentation, treat acne, or manage conditions such as melasma (see pages 76–77) or actinic keratoses. Some people use them for their rejuvenating effects, others are guided by a dermatologist to address specific concerns.

What actually happens?

The process involves applying a chemical solution that temporarily disrupts the bonds between skin cells, allowing the uppermost layers to shed. The deeper you go with a peel the more dramatic the outcomes are in terms of skin texture and tone, but this also comes with a longer recovery period while the skin fully heals.

Types of chemical peel

Peels vary in depth – superficial, medium, or deep – depending on the agent used and the condition being treated.

Superficial peels, such as those using alpha hydroxy acids (glycolic or lactic acid), target the outermost layer of skin and are commonly used to brighten dull skin, treat mild pigmentation, and manage acne. These require little to no downtime and can be performed regularly.

Medium-depth peels, often using trichloroacetic acid (TCA), penetrate further into the epidermis and upper dermis. They are effective for treating sun damage, more pronounced pigmentation, and mild scarring. Recovery may take several days, with redness and peeling.

Deep peels, such as phenol-based formulations, reach the deeper dermis and are reserved for more severe wrinkles, precancerous growths, and significant photo-ageing. Deep peels are typically a one-off procedure and require a longer recovery – one to two weeks of visible healing, with ongoing redness that may persist for weeks or months.

Generally, chemical peel complications are rare. They correlate with depth of peel – the deeper you go, the higher the risk of scarring, infection, or prolonged pigmentation changes.

-)-)-)-

botulinum toxins

Derived from the bacterium *Clostridium botulinum*, botulinum toxins are among the most recognizable injectable treatments in clinical medicine. Yet their role in the reduction of wrinkles is probably the least interesting use of this versatile protein.

Though best known for its cosmetic effects, Botox is a brand name for one of several formulations of botulinum toxin, a substance that has its roots firmly in medical treatment. Long before it was smoothing foreheads, botulinum toxin was – and still is – used to treat a wide range of medical conditions, including muscle spasms, excessive sweating, migraines, and even certain bladder and gastrointestinal disorders. Its ability to safely and temporarily relax overactive muscles has made it a versatile tool in both medicine and aesthetics.

How does the procedure work?

For a muscle to contract, it needs to receive a signal from a nearby nerve ending. That signal comes in the form of a chemical messenger called acetylcholine, which is released by the nerve and picked up by the muscle. Once this happens, the muscle tightens. Botulinum toxins work by quietly interrupting that process. They block the release of acetylcholine at the point where the nerve meets the muscle – no signal means no contraction. The result is a temporary softening or relaxation of the muscle.

The way botulinum toxins do this is by targeting a group of helper proteins known as SNAREs (soluble NSF attachment protein receptors). These proteins allow chemical messengers to be released from nerve cells. By halting the action of the SNARE proteins, the toxins prevent not just acetylcholine, but also other chemicals such as glutamate and substance P from being released. These latter two are involved in pain pathways, and we are slowly understanding more about the role that botulinum toxins may play here.

Once injected, the effects usually begin to show within two to four days, with full muscle relaxation seen by around day 10. These effects wear off over time. The nerves gradually grow new endings and reconnect with the muscle, usually by the four-month mark, after which the nerves start working normally again.

Effects on the skin related to the muscle

Botulinum toxins are used to manage a range of conditions where muscle movement or nerve-driven signals contribute to the problem. These include relaxing facial lines that form with

-)-)-)-

repeated expressions, such as those across the forehead, between the eyebrows, or around the eyes. Their ability to reduce excess sweating (hyperhidrosis) is another well-established use, particularly for the underarms, palms, or soles.

Botulinum toxin also affects the skin in ways that do not involve nerves or muscles. That's because skin cells like keratinocytes and immune cells also use acetylcholine. When this is blocked, it can be useful for skin problems such as rosacea (see pages 74–75), oily skin, itching, and scars.

Different types of botulinum toxin

Although all these treatments work in a similar way, different brands have slightly different compositions:

- **Botox** (onabotulinumtoxinA): the original formulation with broad regulatory approval for dermatological and neurological indications.

- **Dysport** (abobotulinumtoxinA): diffuses slightly more, potentially covering a larger treatment area.

- **Xeomin** (incobotulinumtoxinA): free of accessory proteins, which may reduce the risk of antibody formation.

- **Jeuveau** (prabotulinumtoxinA): developed with a focus on neuromodulatory effects.

Botulinum toxin B (Myobloc) is reserved for patients who develop resistance to type A or for specific neurological indications.

HOW BOTULINUM TOXINS RELAX MUSCLE FIBRES

SNARE proteins help the nerve ending fuse with the muscle fibres so the acetylcholine can be released. When botulinum toxin cuts one of these SNARE proteins, the nerves can't release acetylcholine, so the nerve signal to the muscle is blocked and the muscle stays relaxed.

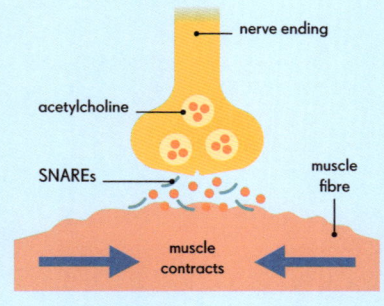

NORMAL MUSCLE FIBRES

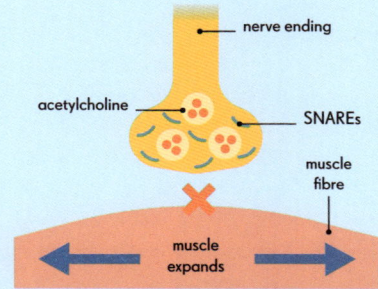

BOTULINUM-AFFECTED MUSCLE FIBRES

fillers

Injectable fillers are substances introduced into the skin
or deeper tissues to restore volume, enhance structural
support, or smooth the appearance of creases and folds.
While many are reversible, others are more permanent.

Hyaluronic acid (HA) fillers are the most widely
used type. HA is a natural sugar in skin and
connective tissue that attracts and holds water,
keeping skin hydrated and elastic. Formulations
vary in thickness and flexibility: softer gels smooth
fine lines or lips, while denser ones add structure
to the cheeks or jawline. The choice depends on
treatment aims and desired longevity. A benefit
is that HA fillers are temporary and reversible.

Alternative fillers

Other fillers work by stimulating collagen
production over time, such as poly-L-lactic acid
(PLLA) and calcium hydroxyapatite (CaHA).
These do not create immediate volume but
gradually firm and thicken the skin. Permanent
fillers like polymethylmethacrylate (PMMA) also
trigger collagen formation but carry greater
long-term risks, including inflammatory nodules.
   When used correctly, fillers can restore volume,
improve scarring, and soften early ageing.
Common side effects include bruising or swelling,
and though rare, injection into a blood vessel can
cause tissue damage. Even with HA, delayed
swelling can occur, so treatment should always
be performed by experienced practitioners.

**HOW A DERMAL FILLER
SMOOTHS WRINKLES**

A filler adds volume beneath the skin,
smoothing wrinkles by plumping and
supporting tissues.

wrinkle • filler injection

epidermis
dermis
hypodermis
muscle

BEFORE

dermal filler

epidermis
dermis
hypodermis
muscle

AFTER

lasers

Since the first medical lasers were used more than
40 years ago, laser technology has advanced dramatically.
It now allows dermatologists to treat a wide range of skin
conditions with far greater precision and safety.

The earliest medical lasers were developed to treat vascular birthmarks such as "port-wine" stains. Though they showed promise, they often left significant collateral damage to surrounding tissue, sometimes leading to scarring.

Laser science is very different today. Using a principle called selective photothermolysis, modern systems can precisely target specific structures in the skin – such as blood vessels, pigment, or water – without harming adjacent cells. This ability to discriminate at microscopic level has transformed lasers into some of the most sophisticated tools in dermatology.

What are lasers used for?

Lasers treat a wide range of dermatological conditions. Scar tissue can be removed, softened, and its colour diluted, while levels of *Cutibacterium acnes* can be reduced to offer a gentle, non-invasive option for mild acne breakouts. Lasers are able to calm inflammation in small patches of psoriasis, and they vaporize stubborn warts and superficial skin lesions.

For the above – as well as for other issues such as tattoo and hair removal – several different types of lasers are used.

VASCULAR LASERS

Vascular lasers work by targeting haemoglobin, the red pigment in blood. They heat the vessel just enough to collapse it, without harming the surrounding skin. Pulsed dye lasers (PDL) and Nd:YAG lasers are two of the most commonly used vascular lasers. PDL is ideal for treating superficial blood vessels such as those in rosacea or "port-wine" stains. It delivers energy in short pulses, heating the vessel wall until it seals shut. Nd:YAG lasers, by contrast, can penetrate deeper to treat larger, deeper vessels – such as leg veins – while bypassing the upper layers of skin.

PIGMENT-TARGETING LASERS

Pigment-targeting lasers seek out melanin, heating the pigment particles until they fragment into smaller pieces that the body can gradually clear away. This same principle can be used to fade brown/black birthmarks and to remove unwanted tattoos. Q-switched lasers, which deliver energy in billionths of a second, were once the gold standard – they shatter pigment with powerful bursts while limiting heat spread. More recently, picosecond lasers that fire even faster pulses have refined the process further. Their ultra-short pulses generate shock waves that break down pigment with less thermal damage, with fewer sessions and better outcomes for stubborn or multicoloured tattoos.

ABLATIVE AND NON-ABLATIVE RESURFACING LASERS

Both these lasers remove the skin, but they do so in very different ways. Ablative lasers vaporize the top layers of skin with extreme precision. This stimulates new collagen and elastin as the skin heals, improving texture, scarring, and deep wrinkles. The results can be dramatic, but downtime and aftercare are more intensive than with non-ablative lasers. The latter leave the skin surface intact, delivering heat into the deeper dermis to trigger collagen remodelling from within. Recovery is quicker, but multiple sessions are usually needed for full effect.

Hair removal

Hair removal lasers work by targeting melanin in the hair follicle, delivering bursts of light that convert to heat and disable the follicle's ability to regrow. The surrounding skin remains unharmed because the laser energy is selectively absorbed by the darker pigment in the hair. Alexandrite lasers are often used for lighter skin types, while Nd:YAG lasers penetrate deeper and are safer for darker skin, as they bypass the melanin in the epidermis. Diode lasers offer a versatile middle ground, with longer built-in cooling for comfort. Generally, results are most effective on coarse, dark hairs, and because hair grows in cycles, multiple treatments are needed.

THE LASER HAIR REMOVAL PROCEDURE

Pulses of laser light are absorbed by the pigment in hair follicles, converting light energy into heat that disables the follicle and prevents future growth while leaving nearby skin unharmed.

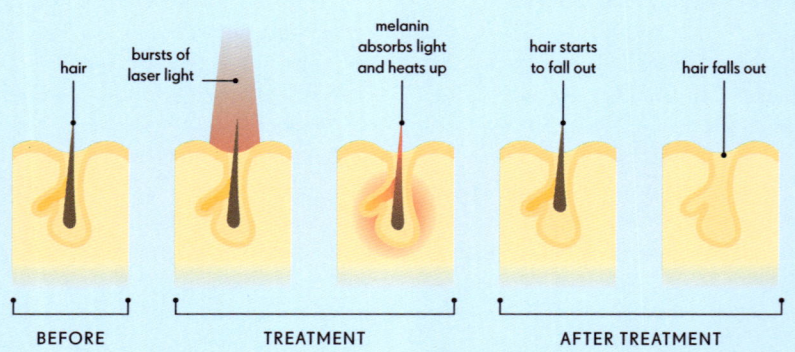

hair

bursts of laser light

melanin absorbs light and heats up

hair starts to fall out

hair falls out

BEFORE   **TREATMENT**   **AFTER TREATMENT**

PRP, stem cells, and exosomes

New skin technologies promise to use your body's own biology to rewind time. Yet, as with many things in aesthetic medicine, the marketing has far outpaced the science.

In clinic, I'm increasingly being asked about the latest regenerative therapies. Let's examine some of the claims behind three new technologies.

PRP

Platelet-rich plasma (PRP) involves taking a small sample of your blood, spinning it in a machine, and then injecting the plasma (now enriched with platelets) back into your skin or scalp. Platelets release growth factors that may stimulate collagen and blood vessel formation. There is some modest evidence supporting its use in early male or female pattern hair loss (see pages 82–83), but very little to support it as a stand-alone treatment for skin ageing or scarring. And not all PRP is created equal – clinics prepare it in different ways, so comparing results is difficult.

Stem cells

Unlike most cells in the body, which have a set role, stem cells are special "starter" cells that can renew themselves and develop into different types of cells. This ability is being explored in several ways. For example, in epidermolysis bullosa (EB), a rare inherited condition where the skin is very fragile, researchers used corrected stem cells from the patient's own body. These were grown into new skin and grafted back, showing what might be possible in the future.

Stem cells are also being studied for their role in wound healing. Those taken from fat or bone marrow may help reduce inflammation, encourage repair, and improve healing in burns or chronic ulcers. In the cosmetic field, stem cell-enriched fat grafts and so-called "stem cell facials" are sometimes used to potentially stimulate collagen and improve skin quality. As yet, there is no firm evidence that these work.

Exosomes

Exosomes are like tiny bubbles released by cells that act as messengers carrying proteins, lipids, and genetic material to influence the behaviour of other cells. Stem cell-derived exosomes are of greatest interest in dermatology. They appear to carry many of the regenerative benefits of stem cells without the risks associated with live-cell therapy. Exosomes are being explored for their ability to promote collagen production, reduce inflammation, and support skin repair. While they are now being added to facials, serums, and post-procedure treatments, proven evidence of their effectiveness in humans is limited.

-)-)-)-

HOW SHOULD I PREPARE FOR A DERMATOLOGY APPOINTMENT?

The key is to make it as easy as possible for us clinicians to see the real story of your skin. A few days before your appointment, jot down your main concerns in order of importance, bring a list or clear photos of any medicines and supplements you are taking, and if you regularly use topical creams, shampoos, or make-up, either bring them or take pictures of the labels; half the diagnosis sometimes lives in the bathroom cabinet. If your problem flares and settles, snap photos on your phone when it is at its worst and keep these in a folder so it is easy to find all of them in the appointment. On the day itself, keep things simple: avoid heavy make-up or thick fake tan on areas you want examined. Wear clothes and underwear that are easy to take off if we need to examine more than one area, and do let us know at the start if there are parts of your body you are not comfortable having examined so we can plan around that. If you are anxious, it is perfectly reasonable to bring someone with you for support, and to arrive with a couple of questions written down. A good dermatology appointment should feel like a focused, honest conversation about how your skin is affecting your life – you should leave having all of your questions answered and with a plan you understand.

—

DOES A SKIN BIOPSY HURT?

For most people, a skin biopsy is more uncomfortable than truly painful – the part you feel most is actually the injection of local anaesthetic rather than the biopsy itself. We start by numbing the area with a tiny needle and a small volume of anaesthetic, which stings for a few seconds, then quickly fades. Once the skin is numb, you should not feel sharp pain, just pressure, pushing, and a sort of tugging sensation if we are doing a punch or excision biopsy. The wound will leave a small mark, but we choose the site and technique carefully to minimize scarring, especially on the face. If at any point during the procedure you feel more than pressure, we can stop and add more anaesthetic – you are not expected to grit your teeth through it!

—

WHAT IS THE DIFFERENCE BETWEEN A DERMATOLOGIST, A COSMETIC DOCTOR, AND A BEAUTY CLINICIAN?

A dermatologist is a medically trained specialist in diseases of the skin, hair, and nails. A cosmetic doctor is not a protected title in most places and the term can be confusing – some are highly trained doctors with extra expertise in aesthetics, while others have far less training in skin disease and systemic medicine. A beauty clinician is usually focused on appearance rather than diagnosis and long-term management of disease. A consultant dermatologist in the UK has to go through full medical school, foundation training, then many years of specialist dermatology training and exams. That covers everything from eczema, acne, psoriasis, and hair loss to autoimmune blistering disease, skin infections, skin cancer surgery, and complex drugs that affect the immune system. A dermatologist can investigate, biopsy, diagnose, prescribe potent medicines, manage complications, and coordinate care with your GP or hospital team, as well as offer aesthetic treatments such as injectables, lasers, and peels.

—

WHAT IF I DON'T AGREE WITH THE DERMATOLOGIST'S PLAN?

If something does not sit right with you, the first step is to say so as plainly as you can in the room: "I'm worried about side effects", "That feels too strong for me right now", "I won't manage something that complicated every day", or "Can we talk about other options?" are all perfectly reasonable responses to the treatment being proposed. That gives your dermatologist a chance to explain why they are suggesting this particular route, what the alternatives are, how long you would need to try it before judging it, and what the plan B and plan C might be if it does not suit you. Often there is more flexibility than it appears at first – lower doses, slower step-ups, different formulations, more monitoring, or starting with skincare and lifestyle changes alongside milder prescriptions. The important thing is that you should never feel trapped in a treatment you do not understand or cannot realistically follow; the best plan is always the one that is medically sound and that you can actually live with.

—

08

Common concerns

dandruff

Dandruff is very common, and while it's rarely serious, it can feel embarrassing or uncomfortable. It isn't a sign of poor hygiene, but a mild inflammatory condition of the scalp known as seborrhoeic dermatitis.

Dandruff tends to flare in areas rich in oil glands (scalp, brows, sides of the nose) and is partly driven by an overreaction to the *Malassezia* yeast that lives naturally on our skin.

You may notice flaking, scaling, or redness around the front hairline or near the ears, and sometimes there is itching. It can worsen in cold weather, with stress, or after illness. The goal isn't to "cure" it permanently, but to manage it well so that it stays under control.

How can I control my dandruff?

I typically advise to start with a medicated shampoo. These are often much more effective than over-the-counter "anti-dandruff" shampoos. Look for active ingredients such as:

- **Ketoconazole** – antifungal
- **Zinc pyrithione** – antimicrobial and anti-inflammatory
- **Selenium sulphide** – slows skin turnover
- **Coal tar** – reduces scaling (but smells strong)
- **Salicylic acid** – helps lift and loosen flakes

Use a medicated shampoo twice a week. Do not apply it like a normal shampoo, but think of it as a scalp treatment – don't wash your hair with it!

Leave it on the scalp for three to five minutes before rinsing. You can rotate between two different shampoo types if needed.

If the dandruff extends to the face (around the eyebrows, creases of the nose, or ears), you can actually use the same shampoo as a face wash two to three times a week – again, leave it on for a minute before rinsing.

When things flare significantly or there's visible redness or irritation, a short course of a mild topical steroid or an antifungal cream can help.

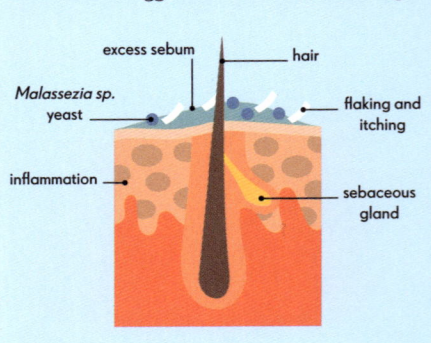

THE FORMATION OF DANDRUFF

Yeasts on the scalp break down oils, creating irritants that trigger inflammation and flaking.

excess sebum — hair

Malassezia sp. yeast — flaking and itching

inflammation — sebaceous gland

folliculitis

Folliculitis is inflammation of the hair follicles. Often mistaken for acne, it can look like a crop of red or pus-filled spots, and tends to show up in areas where friction, heat, or shaving disrupt the skin — the underarms, thighs, buttocks, and the beard area.

Folliculitis is usually caused by bacteria, most commonly *Staphylococcus aureus*, but can also result from yeast or even irritation from sweat and friction. Some cases are linked to ingrown hairs, particularly in curly or coarse hair types. In people who frequently wax, shave, or wear tight clothing – especially gym clothes – folliculitis can become recurrent.

Treatment

The key is to treat both the infection (if present) and the inflammation. For mild cases, a topical antiseptic wash – such as 4% chlorhexidine, benzalkonium chloride, or benzoyl peroxide – used daily for seven days can settle things down. Avoid shaving or waxing while the skin is inflamed. If there's no improvement, or it worsens, a topical or oral antibiotic may be needed. If the infection requires antibiotics, it is advisable to take a skin swab beforehand so the right medication can be selected based on how sensitive the bacteria is to antibiotics available.

Fungal folliculitis (caused by yeast) looks similar, but often itches more and doesn't respond to antibiotics. In fact, in some cases this is actually caused by being on antibiotics. That's where a topical or oral antifungal becomes important and a doctor can guide you.

Prevention

Preventing folliculitis in the long term often means making a few practical changes. If you remove hair, consider switching to a gentler method – shaving in the direction of hair growth with a clean razor or laser hair reduction (see page 156), which can dramatically reduce the problem. Minimize friction from tight clothing, and keep areas prone to heat and sweat clean and dry.

For those who experience recurrent episodes, it may be helpful to take a skin swab to check for bacterial carriage, particularly *Staphylococcus aureus*. Using an antiseptic wash such as chlorhexidine or benzoyl peroxide once a week can help reduce flare-ups.

-)-)-)-

cold sores

If you've had a cold sore, you'll recognize the telltale tingle before the blister appears, usually in the same place every time.

Cold sores are caused by herpes simplex virus type 1 (HSV-1). It is a virus that lies dormant in the nerves and can reactivate from time to time, often triggered by stress, illness, sunlight, or hormonal shifts. While most people associate HSV-1 with the lips, it can appear in other places too – where the virus shows up is where it entered the body in the first place. The area around the mouth remains the most common site, which is why we think of it as the "cold sore" zone. Wherever it appears, the pattern is usually the same: a tingle or burn, followed by small, fluid-filled blisters that crust over as they heal.

Therapies and triggers

Treatment works best when you catch it early. As soon as you feel the tingling, apply a topical antiviral cream (such as aciclovir) several times a day. This can reduce the duration of the outbreak and make it less severe. For people with frequent or severe flare-ups, oral antiviral tablets may be more effective and are often best taken at the first sign of symptoms.

Sunlight is a common trigger, so use a lip balm with SPF30 or above every day, especially if you're skiing, sunbathing, or travelling. Avoid kissing or sharing utensils during an outbreak, as cold sores are highly contagious until they're completely healed. They can generate quite a serious reaction if you pass the virus to a person (usually a baby) who has not been exposed to it before.

Don't pick at the blister or peel off the scab, as this can lead to delayed healing or scarring. Keep the area clean and moisturized with a bland lip balm or barrier ointment. My patients usually find it very helpful to place a small circular hydrocolloid dressing over their cold sore for therapeutic and cosmetic benefit – you can buy these in the form of "zit stickers" or "pimple patches".

If your cold sores are becoming frequent (more than six episodes a year), consider preventive antiviral medication, especially if the sores are impacting your quality of life.

-)-)-)-

chapped lips

Chapped lips, or cheilitis, may seem like a minor nuisance,
but for some people it can result in lips that are persistently
sore and even crack or bleed. Let's find out why.

Your lips are particularly vulnerable because they lack oil glands and have a thinner protective barrier than the rest of the face. They become chapped because of a loss of water through their surface, leaving them prone to cracking. As we move our lips so much, this can lead to bleeding.

What sets off chapped lips?

The most common form of the condition is irritant cheilitis, which is triggered by dry air, cold weather, wind, licking your lips, and fragranced or mentholated balms. Ironically, lip balms can make things worse by causing irritation or dependency – the more you use them, the drier your lips feel. Also, many include menthol and other fragrances that make the problem worse – your lips might initially feel better after applying the balm, but often it can just fuel the problem.

Sometimes cheilitis is linked to eczema, irritation, or allergy to lip products. Angular cheilitis affects the corners of the mouth and is often caused by saliva irritation. Actinic cheilitis develops from long-term sun exposure, causing dryness or scaling, and is considered precancerous. Granulomatous cheilitis is rare and causes firm swelling from deeper inflammation.

Keep lip care simple. Avoid flavoured or scented balms and use plain, fragrance-free products with

occlusives such as petrolatum. If lips are cracked, apply a healing ointment after gentle exfoliation once or twice a week. A short course of hydrocortisone 1% ointment can help, but prolonged use may cause perioral dermatitis. Persistent soreness at the corners of the mouth usually responds to a mild antifungal-steroid cream or a baby cream with titanium dioxide.

SELECTED TYPES OF CHEILITIS
The discomfort and cosmetic distress of cheilitis can significantly affect daily life.

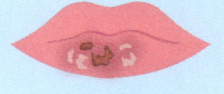

actinic cheilitis

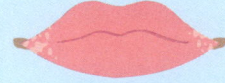

angular cheilitis

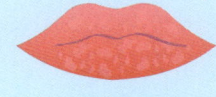

granulomatous
cheilitis

heat rash

A heat rash develops when your sweat glands get a bit overwhelmed, causing sweat to get trapped under the skin. It is common in babies, particularly in warm environments or if they have had a fever, but adults can also get the rash.

Heat rash, or miliaria, is something I see quite often, especially during warmer months or in humid climates. While anyone can get it, some people are more prone than others. Babies are particularly susceptible because their sweat ducts are still developing.

Heat rash is especially common in young infants. It presents as a red, bumpy rash, particularly when the infant has had a lot of skin-to-skin contact and cuddles while their sweat gland openings are maturing. It is very common, and the cause of most telephone calls I get from friends with a new baby. In adults, it's more common in those who sweat heavily, wear tight clothing, or have limited mobility.

The mildest form of heat rash is called miliaria crystallina. It occurs when the opening of the sweat duct on the surface of the skin (sweat pore) is blocked. This form is marked by tiny, clear, fluid-filled bumps that break easily.

Another type that occurs deeper in the skin is known as miliaria rubra, which is sometimes called prickly heat. Signs and symptoms include small, inflamed blister-like bumps and itching or prickling in the affected area.

How can I avoid getting it?

First off, prevention really is the best medicine here. Heat rash tends to show up when skin gets hot and sweaty, so your goal is to keep cool and dry as much as possible. I always recommend wearing loose, breathable clothes – soft cotton rather than synthetic fabrics – because tight clothes can trap heat and irritate your skin further.

Keeping cool is crucial. Try to stay in air-conditioned spaces or use a fan, especially if you're active or outside in the heat. Take regular breaks to cool down and avoid sitting in direct sunlight for long periods.

Treating heat rash

Wash the area gently with cool water. After washing, pat your skin dry rather than rubbing it, and if you can, let your skin air-dry before dressing.

For relief, over-the-counter options such as a mild hydrocortisone cream can soothe the itching and inflammation. Just be careful to avoid heavy creams or oily products that might block your sweat glands even more.

-)-)-)-

sunburn

Sunburn is the result of your skin being exposed to too much ultraviolet (UVB) light, causing redness, pain, and sometimes peeling. Let's take a look at how to prevent and treat it.

Sunburn should be avoided. Research shows that having five or more burns doubles your risk of getting skin cancer. Even one blistering sunburn in childhood more than doubles your chances of later developing melanoma.

The key is to try and prevent this from happening. The UV index is a simple but powerful guide to help you judge the risk of sunburn on any given day. It reflects the strength of UVB rays (see page 94) – the main culprits behind burning – and ranges from 0 to 11+. When the index rises above 3, extra protection is advised. Even on cooler or overcast days, a high UV index can mean your skin is still at risk.

When the UV index is high, the best approach is layered protection. Broad-brimmed hats shield the face, ears, and neck, while lightweight, long-sleeved clothing offers a physical barrier.

Tips to relieve sunburn

The key is to soothe your skin while it heals. Get out of the sun, cool down, and follow these tips.

- Cool your skin with a clean towel dampened with cold water, or take cool baths. Drink plenty of water to prevent dehydration.

- Apply soothing moisturizers such as aloe vera or calamine lotion to tender or peeling skin.

- Avoid popping blisters – intact blisters protect healing skin. If one breaks, clean the area, apply antibiotic cream, and cover with a bandage.

- For mild/moderate sunburn, apply 1% hydrocortisone cream up to three times a day. Over-the-counter pain relievers can reduce pain.

THE UV INDEX

The UV scale measures the intensity of ultraviolet radiation, acting as a guide to sun protection. Note that the scale changes through the day and differs depending on where you are in the world.

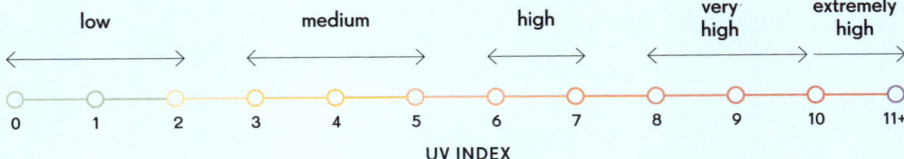

| low | medium | high | very high | extremely high |

| 0 | 1 | 2 | 3 | 4 | 5 | 6 | 7 | 8 | 9 | 10 | 11+ |

UV INDEX

-)-)-)-

broken blood vessels

Tiny, visible blood vessels on the face or legs are common and usually harmless. Often just a sign of ageing, sun exposure, or genetics, they can sometimes hint at underlying skin or health conditions — but most of the time, they're nothing to worry about.

For most people, broken blood vessels, or telangiectasia, are a harmless and purely cosmetic concern. Occasionally, they can point to something more – increased vascular fragility, early signs of rosacea (see pages 74–75), cumulative UV damage (see pages 94–95), or poor circulation in the legs. It's rarely serious, but it's worth understanding what your skin might be trying to tell you.

Broken vessels on the face most frequently develop around the nose, cheeks, and chin. They're usually caused by factors such as sun damage, genetics, hormonal changes, or even triggers such as wind and extreme temperatures.

On the legs, these vessels often appear as a result of increased pressure in the veins. This happens when blood doesn't flow back to the heart efficiently. The extra pressure causes tiny surface vessels to widen and become visible over time. Standing for long periods, pregnancy, or hereditary vein problems can all contribute.

Evaluating the situation

It's important to recognize not only the appearance of these vessels but also any symptoms such as aching, heaviness, or swelling in the legs, which can indicate deeper vein issues.

ON THE FACE

My treatment of choice for the face is usually laser therapy (see pages 155–56). I use a vascular or pulsed dye laser, which targets the haemoglobin (oxygen-carrying protein) in these tiny vessels, causing them to collapse and fade over several sessions. The procedure is quick, minimally uncomfortable, and requires little downtime. If patients have rosacea, I combine laser sessions with topical treatment to reduce inflammation.

ON THE LEGS

Treating veins on the legs can be more involved as the vessels are usually deeper. For small, surface-level veins, sclerotherapy – where a solution is injected into the vein to seal it shut – is often effective. Before that, it's important to check whether deeper veins are under strain. If blood isn't flowing properly back up the legs, the pressure can keep creating new vessels. Here, a duplex ultrasound scan can help decide the next steps. Larger veins may first need surgery or treatment with a laser inside the vein wall.

Alongside any procedure, you should stay active, elevate your legs, and wear compression stockings if you're on your feet for long periods.

-)-)-)-
.:.:.:.

keratosis pilaris

A surprising number of people have keratosis pilaris (KP) – it affects up to 70 per cent of teenagers and 40 per cent of adults. The cause is not fully understood, but it is genetically linked and often inherited from a parent.

KP occurs when keratin builds up and blocks the opening of the upper hair follicle instead of shedding normally. This leads to small, rough bumps often likened to "chicken skin".

The condition typically affects the outer upper arms and thighs, but may also appear in other areas. It can present with redness (keratosis pilaris rubra), bumps on the face, or thinned eyebrows.

KP is worse in winter when the air is dry, and better in summer when there's more humidity. In stubborn cases, we sometimes use topical retinoids or laser treatments to reduce persistent redness or roughness, especially on the face.

Treatment options

There's no cure for KP, and it is not a condition you can scrub away, but you can soften its appearance with the right ingredients.

Urea (5–10%) – Yes, urea is found in urine, but the kind in your cream is made in a lab. It helps break down the hardened keratin that clogs follicles and draws moisture into the skin at the same time.

Lactic acid – These exfoliating acids dissolve the bonds between dead skin cells, making it easier for them to shed without irritation.

Ceramides and glycerin – These support your skin barrier, locking in moisture to smooth the skin.

HOW KERATOSIS PILARIS DEVELOPS
Keratin-filled corneocyctes (see page 18) clog the follicle, causing irritation and skin bumps.

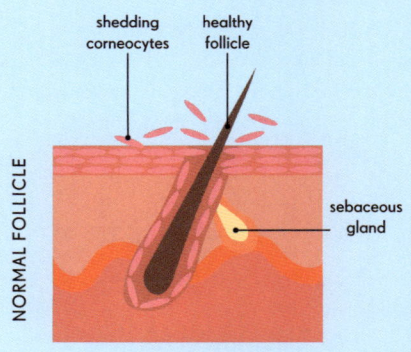

NORMAL FOLLICLE

shedding corneocytes

healthy follicle

sebaceous gland

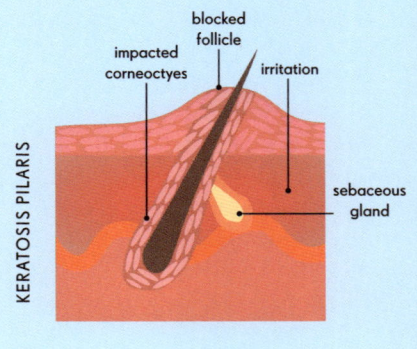

KERATOSIS PILARIS

blocked follicle

impacted corneoctyes

irritation

sebaceous gland

-)-)-)-

athlete's foot

When we talk about athlete's foot, we're actually talking about a fungal infection on the foot. While the condition can cause some discomfort, there are measures you can take to treat and prevent it.

The main culprits behind athlete's foot are a group of fungi known as dermatophytes. These organisms are specialized in digesting keratin, and they thrive in warm, damp conditions. They don't invade beyond the outermost layers of skin, but they can cause significant irritation, inflammation, and scaling. The medical term for athlete's foot is tinea pedis, and depending on what site of the body is involved, there is usually a fun term to go alongside it.

How do I know if I have it?

Athlete's foot usually starts between the toes, particularly the fourth and fifth digits, where the skin becomes itchy, cracked, white, and soggy-looking. From there, it can spread to the soles or sides of the feet, where you might notice dry, scaly, or even blistered patches. Many people mistake the condition for dry skin or eczema, which is why it's often left untreated or mismanaged. If it gets out of control, it can spread up the leg or you can even develop an itchy rash on other parts of the body as the immune system reacts to the fungus.

The most common dermatophyte I see is *Trichophyton rubrum*, while another frequent offender is *Trichophyton interdigitale*, which tends to cause the classic soggy, peeling appearance between the toes. Less commonly, *Epidermophyton floccosum* may be the culprit – it's another keratin-loving fungus that causes similar symptoms, but it's most prevalent in tropical regions.

Treatment and prevention

What I typically recommend is topical antifungal treatment as a first step, something with terbinafine, clotrimazole, or miconazole applied twice daily to clean, dry feet. You need to keep using it for at least a week after the skin looks normal, or the infection will likely come back. For more stubborn or widespread cases, especially if the soles are thickened and scaling (a pattern called moccasin-type tinea), I may prescribe a short course of oral antifungals.

If you are prone to athlete's foot, prevention is important. Keep your feet dry, especially between the toes – you can use talc to absorb any additional moisture. Change socks daily, use a separate towel for your feet, and if you're using communal showers or pool areas, always wear flip-flops.

-)-)-)-

scabies and bed bugs

Mites are the source of scabies, while bed bugs are small insects that infest your home and irritate your skin. Let's examine the role these two members of the animal kingdom play in common but uncomfortable skin conditions.

Scabies

Scabies is an intensely itchy, distressing, and very contagious condition. It's caused by the *Sarcoptes scabiei* mite, which burrows into the stratum corneum to lay its eggs.

The mites are transmitted by prolonged, direct skin-to-skin contact with an infested individual, typically lasting at least 5–10 minutes or more.

Everyone in the same household – and anyone with regular skin-to-skin contact – should be treated as if they've been exposed, even if they don't have symptoms. The rash can take several weeks to show up, and if just one person is left untreated, the whole cycle can start again. Treatment usually starts with permethrin 5% cream, applied to the entire body, including under the nails, around the genitals, and between the fingers and toes. This is repeated seven days later to catch any newly hatched mites.

In more severe cases, or when the skin is covered with thick crusts and flakes, we often add a tablet called ivermectin to help clear the infection.

If you or anyone in your household has scabies, all bedding and clothing should be hot-washed. The itch may persist after the mites are gone, so topical steroids can ease the inflammation.

Bed bugs

Bed bugs (*Cimex lectularius*) are tiny, flat, wingless insects that feed on human blood, typically at night. They don't live on the skin, but hide in mattresses, bed frames, and cracks in furniture. Their bites often appear in a line or cluster and can cause very red, itchy welts. The treatment of bed bug bites is based around topical steroids and antihistamines.

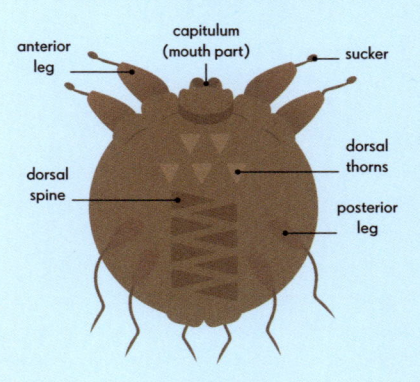

ANATOMY OF A SCABIES MITE

A female adult scabies mite is only 0.3 mm (0.01 in) long, so almost invisible to the eye.

anterior leg

capitulum (mouth part)

sucker

dorsal thorns

dorsal spine

posterior leg

-)-)-)-

unwanted facial hair

From a few stray chin hairs to more widespread facial hair growth related to hormonal change, unwanted facial hair is an issue that can deeply affect confidence and self-image. Nevertheless, there are several effective options for managing it.

Unwanted hair can be a normal part of ageing, especially around hormonal shifts like perimenopause. Yet hirsutism – the growth of coarse, dark hair in a pattern more typical of male hair growth – is often driven by androgens, the so-called male hormones that all of us have to varying degrees. An increase in sensitivity to these hormones, or higher levels in general, can lead to more noticeable hair growth. This is common in conditions such as polycystic ovary syndrome (PCOS), but can also be linked to genetics, certain medications, and rarely, underlying conditions like congenital adrenal hyperplasia.

If the hair growth is sudden, rapidly increasing, or accompanied by acne, irregular periods, or other hormonal signs, it's worth checking in with a doctor.

Simple methods like shaving, waxing, or threading are widely used because they're quick. Shaving doesn't make hair grow back thicker or darker – it just feels coarser because of the blunt edge. Depilatory creams dissolve surface hair chemically, but they can irritate sensitive skin and should be patch-tested first (see page 143).

Traditional methods such as sugaring (using a natural paste) and threading (using a twisted thread to remove hairs) can be very effective. Another option for small areas or lighter hairs is electrolysis, where a fine probe destroys the hair follicle with electrical current. It offers permanent results, though it takes some time.

Ultimately, it's not about whether you remove the hair, but about finding the approach that fits your skin, your preferences, and your values.

Managing unwanted hair

Unwanted hair is common, and how we choose to manage it – or not – is deeply personal. For some, it's just the occasional chin hair that crops up with age; for others, especially those with hormonal conditions such as PCOS, it can involve thicker or more widespread hair growth. Genetics also play a role, and what's considered "unwanted" can vary in cultures and individuals.

Laser hair removal

For larger or denser areas, laser hair removal (see page 156) is the real game changer. It works by targeting melanin within the hair follicle with focused light energy, damaging the follicle to reduce regrowth over time. The best results are seen in those with dark hair on lighter skin, as the contrast helps the laser zero in on the follicle while sparing the surrounding skin.

-)-)-)-

dark eye circles

If there's one area that gets more attention than it deserves it's the under-eye, and the truth is that not all "dark circles" are created equal. Understanding *why* they're there is the key to treating them.

There are several reasons why people develop pigmentation or shadowing beneath the eyes. Anatomically, the skin under the eyes is the thinnest on the body, so blue veins can appear as dark shadows to create vascular circles. If someone has a deep tear trough – the hollow between the inner corner of the eye and the cheek – it creates a shadow that exaggerates the darkness, especially under overhead lighting.

Structurally, the position of the globe of the eye in the orbital socket and the volume of the fat pads around the lower eyelid also influence how light hits the area. Plus, as we age, the fat supporting the under-eye region can shift or reduce, worsening these shadows.

The role of pigmentation

In skin of colour, it's quite common to have naturally higher melanin in the under-eye area. In addition, anything that causes friction or inflammation, like frequent eye-rubbing from allergies, eczema, or hay fever, can lead to darker pigmentation over time. Lifestyle matters, too: poor sleep, dehydration, smoking, and stress all affect the tiny vessels under your eyes.

In terms of treatments, a good moisturizer containing hyaluronic acids, ceramides, and glycerine will help. Topical lightening agents, laser therapies (see pages 155–56), and chemical peels (see page 151) can improve pigment, while tear-trough filler can help volume-related shadows.

TYPES OF DARK EYE CIRCLES
Whatever variety of dark circles you may have under your eyes, the most practical daily option to lessen their impact is to camouflage with colour-correcting concealer.

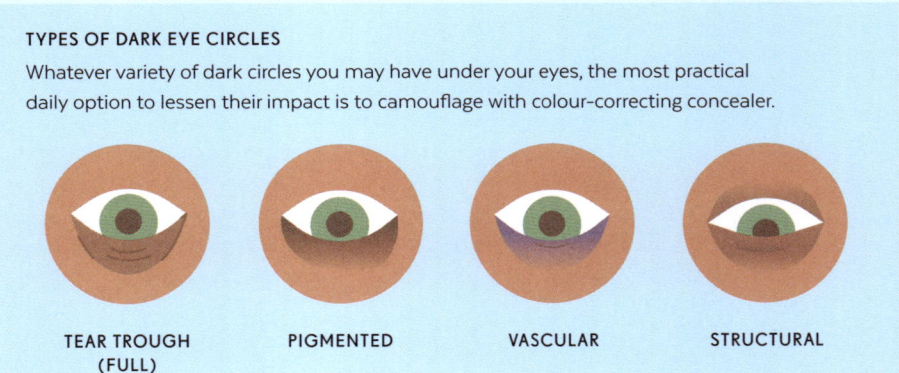

TEAR TROUGH   PIGMENTED   VASCULAR   STRUCTURAL
(FULL)

HOW SHOULD I DEAL WITH A SMALL CUT ON THE SKIN?

For a fresh cut, the goal is very simple – clean, close, and protect. First, gently rinse the area under cool or lukewarm running water; if there is visible dirt, use a mild, unfragranced soap around the wound. If it is bleeding, apply firm, direct pressure with a clean cloth or dressing for 5–10 minutes. Most small cuts will slow and then stop. If blood is spurting, soaking through dressings, or you are on blood thinners and cannot get the bleeding under control, you need urgent medical attention. Once the wound is clean and the bleeding has stopped, you do not need to add anything advanced to the site – just a thin smear of plain petrolatum or a simple antiseptic, followed by a non-stick dressing to keep the wound moist and covered. Moist healing gives you a better cosmetic result than letting it "air out" and scab; the traditional advice to leave wounds open to the air has largely been dropped. Change the dressing once a day, and reapply the ointment. Once the cut has fully closed, you can switch to simple moisturizer and, if the wound is in a sun-exposed area, daily SPF – early sun protection on a new scar is one of the best ways to help it fade as neatly as your skin will allow.

———

WHAT IS THE BEST WAY TO MANAGE A BURN?

For a thermal burn, the most important first step is to run cool or lukewarm running tap water over the area as soon as you can and keep it there for around 20 minutes. This helps limit how deep the heat penetrates and can make a real difference to pain and scarring. While you are cooling the burn, gently remove jewellery, watches, or tight clothing near the area before swelling starts. Once the skin has been cooled, look at what you are dealing with. A superficial burn (red, sore, a bit puffy, like bad sunburn) can usually be managed at home. Pat the area dry with something clean, then apply a simple, non-adherent dressing or sterile gauze with a thin layer of plain ointment (such as petrolatum) underneath to keep it moist; moist wound-healing gives a better cosmetic result than letting it dry, crack, and scab. If there are blisters, the burn is tender even at the edges, or it covers a large area, you should get the burn assessed by a clinician.

———

HOW DO WARTS GETS PASSED AROUND?

Warts are little viral souvenirs your skin picks up from other people or from the environment. The HPV virus that causes warts is passed on when it meets skin that has tiny breaks in the barrier, especially soggy or wrinkled skin from swimming pools. It can be transmitted through direct skin-to-skin contact or indirect contact via damp surfaces and objects where the virus has been left behind, such as poolside tiles, communal showers, or gym mats. Once one area is infected, you can also give warts to yourself – picking at warts, shaving over them, or vigorously pumicing the feet can move the virus along the skin so new lesions pop up nearby or along lines of trauma. The incubation period is long, so people rarely remember the exact moment they caught the virus.

—

WHAT HAPPENS WHEN YOU GET A COLD SORE FOR THE FIRST TIME?

This is your first proper encounter with herpes simplex virus type 1 (HSV-1). For many people, the primary infection is so mild they barely notice it, but in others, the first episode can be much more dramatic – several days of feeling under the weather, swollen gums, painful mouth ulcers, drooling or refusing to eat because everything stings, low-grade fever, and tender glands under the jaw. We call this herpetic gingivostomatitis, and it is one of the commonest reasons for very sore mouths in otherwise healthy children. The lips may crack and crust, and there can be obvious small blisters at the border between the lip and the surrounding skin, which then burst and scab. During this first infection, the virus is actively replicating in the skin and mucosa, and your immune system is working hard to get on top of it. As the episode settles (usually between 7–14 days), the virus does not leave the body. In some people, it will reactivate from time to time, usually when the local skin is stressed (sunburn, chapped lips), the immune system is under extra pressure (illness, menstruation, major stress), or there is direct trauma. The main red flags in a primary episode are if you or your child cannot drink, are very lethargic, or if sores appear near the eye. In these cases, antiviral tablets and an urgent medical review are important.

—

09

Systematic health

the skin as a mirror

The skin is uniquely positioned to alert us to trouble elsewhere in the body. Let's examine how marks, lumps, and bumps can be a reflection of a range of internal conditions.

One of the most fascinating aspects of dermatology is how the skin so often speaks on behalf of the rest of the body. Sometimes it whispers subtle change, other times it shouts with a sudden eruption.

Erythema nodosum

The painful, bruise-like nodules of erythema nodosum tend to cluster on the shins and sometimes the arms, evolving from red to purple to yellow-green. The nodules sit in the fat layer of the skin and can feel warm, swollen, and tender. The condition is associated with a variety of triggers, including streptococcal throat infection, tuberculosis, inflammatory bowel disease (particularly Crohn's), certain medications, and even pregnancy. In many cases, the nodules arrive before the underlying condition becomes clear.

Autoimmune diseases and cancers

Similarly, in autoimmune and rheumatological disease, the skin often leaves clues. Rheumatoid arthritis can produce firm nodules under the skin. Other clues might include livedo reticularis, a pattern of purplish discolouration caused by sluggish blood flow, or leg ulcers from small vessel inflammation. Blood-vessel syndromes often first appear on the skin as painful purple spots that don't blanch with pressure, nodules, or ulcerations. These can be early signs of internal organ issues.

A flushed, butterfly-shaped malar rash across the cheeks and nose can be the first sign of lupus erythematosus (SLE). In dermatomyositis, the skin may develop a heliotrope rash around the eyes or Gottron's papules on the knuckles.

Sometimes, the first sign of a breast or lung cancer is a firm, painless nodule in the skin. Leukaemia and lymphoma, too, can produce unconventional rashes: itchy, persistent, or just strangely unresponsive to normal treatment.

Signs of poor nutrition

The skin also shows us nutritional deficiencies – hair thins, nails become brittle, and skin loses its glow. Iron and B12 deficiency can present as pallor or hair shedding, while a lack of vitamin C (scurvy) brings tiny bleeds around hair follicles and "corkscrew" hairs where defective collagen causes twisted hair shafts. Zinc deficiency can show up as a red rash around the mouth, hands, or genitals, while niacin deficiency (pellagra) causes rough, hyperpigmented skin in sun-exposed areas.

-)-)-)-

SKIN INSIGHT BODY MAP

Typical body distribution of rashes that hint at underlying nutritional or systemic disease.

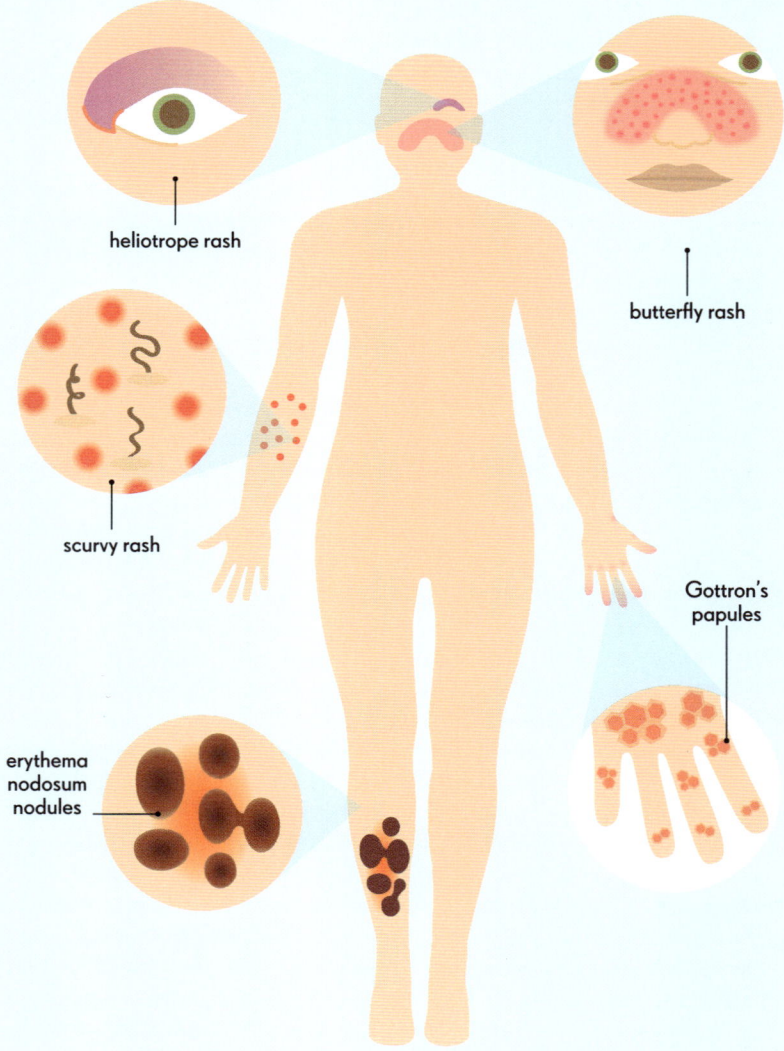

heliotrope rash

butterfly rash

scurvy rash

Gottron's papules

erythema nodosum nodules

the gut-skin axis

What we eat, how we digest, and the microscopic ecosystems
that live inside our gut have direct consequences for the health
and function of our skin. When things go awry in the gut, we
can often spot the signs on the skin.

The once-offbeat idea that your gut and your skin might be talking to each other is now well supported by science. These two organs communicate with each other through the microbiome, immune, and hormone signals. Gut microbes break down food into metabolites such as vitamins and short-chain fatty acids, which enter circulation and influence skin immunity and repair. In return, the skin sends messages back via hormones, vitamin D made from sunlight, and molecules and proteins released after injury.

The gut barrier and microbiome

At the heart of this conversation is the gut barrier. When healthy, it acts like a sieve – absorbing nutrients while keeping out harmful substances. Yet when the barrier is compromised, often described as a "leaky gut", bacteria and toxins can enter the bloodstream to create a condition called dysbiosis. This triggers immune activation that may show up on the skin as redness, itching, or flare-ups. For those with acne, rosacea, psoriasis, or eczema, gut-driven inflammation can influence the timing and intensity of symptoms.

The gut microbiome is incredibly important. The trillions of bacteria, viruses, and fungi living in our intestines ferment dietary fibre to produce short-chain fatty acids such as butyrate, which calm inflammation and support the gut lining.

Gut-influenced skin conditions

Acne sufferers often have reduced diversity in their gut microbiome and fewer butyrate-producing bacteria. Trials have shown that a mix of oral probiotics and antibiotics improves acne and reduces gut issues. A low-glycaemic-load diet, which regulates blood sugar spikes, also reduces acne, possibly by improving insulin sensitivity and creating a healthier gut microbiome.

Rosacea also has a surprising gut connection. It has been linked with several gastrointestinal disorders, including irritable bowel syndrome, inflammatory bowel disease, and small intestinal bacterial overgrowth (SIBO). In fact, SIBO appears to be significantly more common in rosacea patients, who have imbalances between pro- and anti-inflammatory gut bacteria. When oral probiotics – particularly non-histamine-releasing strains – are added to treatment, SIBO patients often report boosts to their skin and wellbeing.

Psoriasis is another condition where the gut appears to play a major role. Sufferers often have

-)-)-)-
·.·.·.

a less diverse gut microbiome, with lower levels of beneficial bacteria. The gut lining may also be more permeable than it should be, allowing inflammatory microbial byproducts to enter the bloodstream. The good news? Dietary shifts – like a Mediterranean-style diet with fibre, polyphenols, and omega-3s – can help dial down inflammation.

The composition of the neonatal gut microbiome appears to influence the development of eczema in infancy. Babies with lower microbial diversity and reduced levels of *Lactobacillus* and *Bifidobacterium* are at greater risk. In older children and adults, improving gutmicrobial health may support eczema control.

A fibre-rich, plant-forward diet with a diversity of vegetables, fruits, wholegrains, and fermented foods such as kefir, kimchi, and sauerkraut supports the growth of beneficial microbes. Probiotic supplements can help, as can reducing alcohol intake and managing stress.

THE MECHANICS OF DYSBIOSIS

Disruption of the gut microbiome alters barrier integrity and immune balance, leading to systemic inflammatory signals that can aggravate skin conditions such as acne, eczema, and psoriasis.

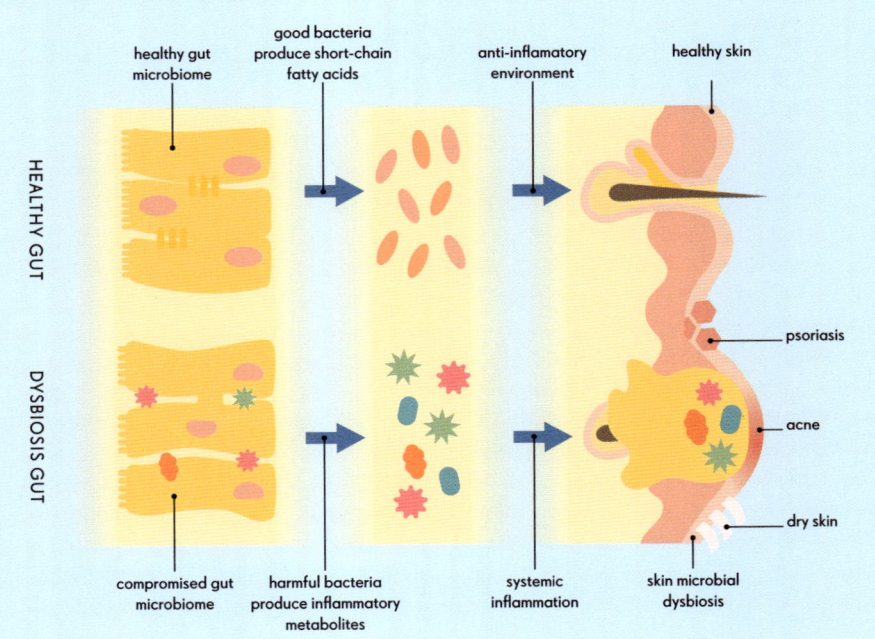

HEALTHY GUT

DYSBIOSIS GUT

healthy gut microbiome

good bacteria produce short-chain fatty acids

anti-inflamatory environment

healthy skin

psoriasis

acne

dry skin

compromised gut microbiome

harmful bacteria produce inflammatory metabolites

systemic inflammation

skin microbial dysbiosis

insulin resistance and type 2 diabetes

Some of the earliest and most telling clues to insulin resistance and diabetes are visible on the skin. When recognized, they can prompt timely intervention and even prevent progression from insulin resistance to type 2 diabetes.

From velvety folds on the back of the neck to clusters of tags under the arms, these skin conditions are often dismissed as cosmetic. In reality, they are powerful biological indicators and often early flags of metabolic disruption.

What is insulin resistance?

The underlying driver here is insulin resistance. When the body's cells stop responding properly to insulin – the hormone that helps move sugar out of the blood – the pancreas compensates by producing more. But insulin doesn't just regulate blood sugar; it can encourage skin cells to multiply and subtly alter how the skin renews and repairs itself. Over time, these internal shifts begin to show on the surface – often well before any lab result signals a problem.

Insulin resistance is one of the things I will often pick up when a patient attends for a skin check of their moles. Almost like a fortune teller, we can see what might happen in the future, so noticing the following signs and suggesting that the patient may need a cholesterol test and a test for diabetes can have an impact on their future health.

Acanthosis nigricans

Let's begin with one of the most striking and under-recognized signs: acanthosis nigricans. This presents as darkened, thickened, velvety plaques that appear in the body's flexural areas, commonly the back of the neck, the underarms, the groin, and sometimes the face. It's a proliferative reaction of the skin to excess circulating insulin. It can appear in children, teenagers, and adults. When it does, it's worth taking a closer look at weight, blood sugar control, and how the body is handling insulin. It's often seen in the context of insulin resistance, particularly in conditions such as obesity and polycystic ovary syndrome, where this metabolic imbalance plays a central role.

Skin tags

Skin tags might seem harmless – and they usually are – but when they start to appear in clusters, especially around the neck, underarms, or groin, they can sometimes be a sign of something more. Several large studies have shown a link between multiple skin tags and raised insulin levels, even in people who don't have diabetes. If they crop up alongside weight gain around the middle, or if there's a family history of type 2 diabetes, it's a gentle nudge to look a little closer at how your body is managing sugar and insulin.

Xanthelasma

Soft, yellowish patches around the eyelids – known as xanthelasma – might seem like just a cosmetic concern, but they can sometimes be a clue that your cholesterol or blood fats are running high. These little plaques are often linked to raised triglycerides or LDL cholesterol, which tend to appear more often in people with insulin resistance. It doesn't mean you have diabetes, but it's a good reason to check your lipid levels – and, in some cases, your blood sugar too.

There is another, more dramatic version called eruptive xanthomata, which presents as small, yellow-red bumps on the buttocks, elbows, or knees. These are rare but important, and they signal that blood fats may be dangerously high and need urgent attention.

Necrobiosis lipoidica

Necrobiosis lipoidica is an uncommon skin condition where shiny, reddish-yellow patches appear – usually on the shins. It's most often seen in people with diabetes, particularly type 1, though it can occur in others too. The cause isn't fully understood, but it's thought to involve changes in the small blood vessels and damage to the skin's collagen. It's more common in women than men

and tends to appear in early adulthood. While the patches can look dramatic, they're not contagious or dangerous.

The role of infections

Infections often appear more frequently when blood sugar is poorly controlled. Fungal infections such as thrush or candidiasis tend to thrive in warm, moist areas of the body – under the breasts, between the thighs, or around the groin. If rashes or soreness in these folds keep coming back, it's worth checking how well your body is managing glucose. The same goes for recurring bacterial infections such as boils or inflamed hair follicles.

Metabolic stress

There are also subtler signs that the skin is under metabolic stress. Delayed wound healing, dry or itchy skin, and increased sensitivity to irritation can all reflect poor microvascular function and the immune system not functioning optimally.

Amazingly, the skin findings don't just reflect internal disease, but also offer feedback during recovery. When insulin sensitivity improves through weight loss, improved diet, regular exercise, and sometimes medications such as metformin or GLP-1 receptor agonists, many of these skin issues begin to resolve: skin tags stop appearing, the dark velvety patches of acanthosis nigricans fade, and inflammatory lesions may settle.

hormones and the skin

Hormones are some of the most influential internal forces
acting on the skin. When their levels rise or fall, the skin
is often one of the first places we see the effects.

Hormones are chemical messengers that travel
through the bloodstream, delivering signals from
glands such as the thyroid, adrenals, ovaries, and
testes. Their influence is far-reaching, shaping
metabolism, mood, and the behaviour of the skin.

The sex hormones

Let's start with the sex hormones – oestrogen,
progesterone, and testosterone.

OESTROGEN

Oestrogen acts as a kind of natural moisturizer
and structural support. It helps the skin hold on
to water, encourages collagen production, and
keeps the deeper layers thick and resilient.
This partly explains the glow many people
notice during pregnancy, when oestrogen levels
rise. But as oestrogen begins to fall – particularly
after menopause (see pages 56–57) – the skin
gradually loses some of that support. It becomes
thinner, drier, and more vulnerable to sun
damage. Supporting the skin through this shift
can involve richer moisturizers, ingredients
that help rebuild the barrier – ceramides or
hyaluronic acid – and, in some cases, hormone
replacement therapy (HRT), which may offer
skin benefits alongside its other effects.

TESTOSTERONE
AND PROGESTERONE

In both men and women, shifts in testosterone
levels can influence body odour, pore size, and
hair; it also has a direct effect on the skin's oil
production. This is why adolescence so often
brings a surge in breakouts, and why conditions
linked to elevated androgens – like polycystic
ovary syndrome (PCOS) – can lead to persistent
adult acne. In PCOS, excess testosterone can
also cause unwanted facial or body hair
(hirsutism) and thinning of the scalp hair.

In men, a significant drop in testosterone over
time may lead to drier, duller skin, less facial hair
growth, and slower skin renewal.

For those experiencing unwanted changes,
treatments range from medical therapy to
cosmetic interventions such as laser hair
removal, retinoids for acne, or procedures
that help restore skin structure.

Progesterone plays a more unpredictable role.
For some, it helps calm inflammation, which
may explain why acne improves in the second
half of the menstrual cycle. For others, it triggers
flare-ups – possibly through its effects on oil
glands and water retention.

-)-)-)-

Thyroid hormones

The thyroid acts like a thermostat for the skin's metabolism. Its hormones – T3 and T4 – help regulate how quickly skin cells renew, how much we sweat, and even how hair and nails grow. When levels drop, as in hypothyroidism, everything slows down: the skin becomes dry, pale, and coarse; the outer edges of the eyebrows may thin; and nails can turn brittle. When levels rise too high, as in hyperthyroidism, the skin often becomes warm and damp, with increased sweating, flushing, and in some cases, hair shedding. These changes are often among the first outward signs of imbalance. While medical treatment is essential, gentle skincare can support the process – rich emollients for dryness, barrier creams for fragile nails, and cooling, non-irritating products for sensitivity and heat. The skin often improves as thyroid levels stabilize.

Cortisol and aldosterone

Then we have the adrenal glands. Perched atop the kidneys, they release cortisol and aldosterone in response to stress, alongside smaller amounts of adrenal androgens. Chronically elevated cortisol, as in Cushing's syndrome, causes the dermis to thin, making the skin fragile and prone to bruising. Stretch marks (see pages 54–55), or striae, may appear wide and reddish-purple in colour, especially over the abdomen, thighs, and upper arms. These deeper-toned marks, known as violaceous striae, result from the skin stretching

when it's lost its normal resilience. There may also be delayed healing and increased facial puffiness. At the opposite end of the spectrum is Addison's disease, where the adrenal glands don't produce enough cortisol. In response, the body tries to compensate by making more ACTH – a hormone that normally tells the adrenals to step up production. Yet when the adrenals can't respond, ACTH keeps rising. This hormone shares a common origin with another called melanocyte-stimulating hormone, which increases pigment production in the skin. That's why one of the telltale signs of Addison's is a deepening of skin tone, particularly in creases, scars, and sun-exposed areas.

Growth hormones

One of the more striking hormonal effects on the skin appears in a condition called acromegaly, where the body produces too much growth hormone – usually due to a benign tumour in the pituitary gland. This leads to high levels of another hormone called IGF-1, which causes the skin to thicken and coarsen, especially on the face, hands, and scalp. In some cases, the scalp develops deep folds and ridges, a change known as cutis verticis gyrata. Other signs include oily skin, large pores, skin tags, increased sweating, and sometimes acne or excess hair growth. The fingers and toes may broaden, and the nails can become thicker. Because these changes come on slowly, they're often mistaken for normal ageing or weight gain.

-)-)-)-

skin and systemic infection

Systemic infections can sometimes whisper their presence before
a fever spikes or a cough sets in. Let's see how viruses, bacteria,
fungi, and the immune system itself can all produce changes
on the skin that hint at something deeper unfolding within.

Even in my own family, I remember one of my children always developing a faint red rash on her thighs a few days before other symptoms of a viral respiratory infection emerged. It was her skin's way of sounding the first alarm.

Viruses

Many of the viral rashes we see in children – measles, rubella, roseola – are classic examples of infections made visible. Measles is usually heralded by fever and cough, followed by a spreading rash, but a very specific clue often comes first: tiny white spots inside the cheeks called Koplik's spots. Rubella tends to be milder, but look out for swollen lymph nodes behind the ears. Parvovirus B19, known for the "slapped cheek" appearance, causes a bright red rash on the face and a lacy eruption on the limbs.

Chickenpox is easy to recognize by its classic mix of spots, blisters, and crusts all appearing at once. Later in life, the virus can reactivate as shingles – a painful, blistering rash that follows the path of a single nerve on one side of the body. There's also a lesser-known condition called Gianotti-Crosti syndrome, often seen in children after a viral illness. It causes a symmetrical rash made up of flat, raised bumps on the cheeks, arms, and legs. Once seen, it's rarely forgotten.

Bacteria

Some of the most telling skin signs belong to bacterial infections. Infective endocarditis, an infection of the heart valves, can show up as painful nodules on the fingers called Osler nodes, as well as painless spots on the palms and soles of the feet known as Janeway lesions. Spotting both together is like cracking a code to an otherwise hidden systemic illness.

Meningococcal sepsis is a rare but serious infection where the skin can offer one of the earliest warnings. It begins when bacteria enter the bloodstream, triggering a powerful immune response that damages blood vessels. Flat, pink spots may quickly darken into purple patches that don't fade when pressed. In severe cases, the skin can lose blood supply and begin to die. It's one of the few rashes that signals a true medical emergency – early recognition and urgent treatment are critical.

Rocky Mountain spotted fever is a serious infection caused by a tick-borne bacterium

-)-)-)-

called *Rickettsia rickettsii*. It starts with a fever, then a rash appears on the wrists and ankles and spreads inwards. The bacteria damage small blood vessels, which is why the rash can darken and become widespread. Early treatment with antibiotics is essential as delays can lead to serious complications.

Scarlet fever often announces itself with a sandpaper-like rash and a bright red, swollen tongue – sometimes called a "strawberry tongue". In infants, another dramatic bacterial illness, staphylococcal scalded skin syndrome, can resemble a burn. Yet instead of heat, it's caused by toxins released from *Staphylococcus aureus* that cause the skin to blister and peel.

Fungal infections

Most superficial fungal infections are harmless and stay on the surface of the skin, but there are rare exceptions. A kerion, which looks like a swollen, inflamed patch on the scalp, can be more severe and may come with fever and swollen glands. In more serious cases, deeper fungal infections such as blastomycosis or cryptococcosis can first appear on the skin. Blastomycosis may cause ulcerated, wart-like lesions that can be mistaken for skin cancer, while cryptococcosis sometimes shows up as small bumps with central dips – subtle signs that may be the first clue to a serious infection, especially in people with weakened immune systems.

Immune response to infection

Sometimes, it's not the microbe itself that causes the rash, but the body's immune reaction to it. Erythema multiforme – a rash made up of target-like spots – is most often triggered by the herpes simplex virus. In children, a more severe version can be caused by *Mycoplasma pneumoniae*, leading to sore skin, mouth ulcers, and other symptoms that affect the whole body.

When bacteria and viruses enter the bloodstream, they spread infection that can often be first deciphered on the skin.

-)-)-)-

cardiovascular health

Your skin often serves as a visible clue to the health of your cardiovascular system. Let's take a look at some of the markers on your skin that all may not be well with your heart.

There is a surprising amount the skin can reveal about the state of our blood vessels and heart. At the same time, some skin diseases do not just reflect cardiovascular risk, but they can also contribute to it.

Xanthomas

One of the clearest skin clues linked to heart health is the appearance of xanthomas – small, yellowish, waxy bumps made up of fat deposits. When they appear on the eyelids, they're called xanthelasma. Around half the people who develop them have raised cholesterol or triglycerides (blood fat).

Another type, called eruptive xanthomata, shows up as sudden clusters of yellow-red spots, often on the elbows, knees, or buttocks. These tend to appear when triglyceride levels spike, usually due to uncontrolled diabetes or inherited cholesterol disorders. If you notice these changes, it's important to check your blood fats and sugar levels straight away.

Arterial disease

Frank's sign, a diagonal crease running across the earlobe, has been statistically linked with coronary artery disease. While it might sound improbable, studies have found that people with this crease may have up to three times the risk of underlying narrowing of the coronary arteries. It is thought to reflect changes in small blood vessel integrity.

Similarly, the skin of the lower legs can become shiny, hairless, and thin in those with peripheral arterial disease (PAD). Here, narrowed leg arteries reduce blood supply, and skin changes can precede more obvious symptoms such as pain when walking (claudication).

Serious heart conditions

Chronic heart failure, or venous insufficiency, commonly results in oedema, the swelling of the lower legs due to fluid build-up. Over time, this can evolve into venous stasis dermatitis, with reddish-brown, itchy, and scaly skin, or progress to lipodermatosclerosis, where the skin and fat of the lower leg become firm and bound down, giving an inverted bottle shape. These changes not only affect quality of life but markedly increase the risk of leg ulceration.

-)-)-)-

Children and the heart

It is important to remember that children can also show skin signs of cardiovascular disease. A good example is erythema marginatum, a red, ring-shaped rash that appears during rheumatic fever, which follows an untreated streptococcal throat infection. Rheumatic fever is a leading cause of heart disease in children globally.

Another example is Kawasaki disease, where children may present with high fever, cracked and bleeding lips, a red tongue, and swelling of the hands and feet. This inflammatory syndrome can cause coronary artery aneurysms if untreated. In both cases, the skin signs are often the first clue to a more serious underlying cardiovascular issue.

Two-way street

The relationship works both ways – some skin conditions can actually raise the risk of heart disease. Chronic inflammatory conditions such as psoriasis don't just affect the skin; they also increase inflammation throughout the body, including in the blood vessels.

People with moderate to severe psoriasis have a higher risk of heart attack and stroke, even if they don't have other common risk factors like high blood pressure or smoking. This is thought to be due to ongoing high levels of inflammatory signals in the blood. Encouragingly, treating psoriasis with systemic or biologic therapies (see pages 146–47) may help reduce this risk while also improving the skin.

CLASSIFICATION OF FRANK'S SIGN EARLOBE CREASE

One interesting clinical association with cardiovascular health is Frank's sign – a diagonal crease on the earlobe observed more often in people with coronary artery disease. The proposed mechanism involves microvascular ageing and loss of elastin fibres, but factors such as age and sun exposure also contribute, making it more a curiosity than a dependable sign.

0
no crease

1
small wrinkle

2a
light crease

2b
crease more
than halfway
across earlobe

3
deep crease
across whole
earlobe

-)-)-)-

the liver-skin connection

The liver is an organ that likes to get on with its
work quietly. But when it starts to struggle,
it is often the skin that raises the alarm.

The liver is busy. It filters the blood, processes hormones, supports clotting, stores vitamins, and builds essential proteins. Sometimes the skin reflects direct signs of liver dysfunction; other times, both organs are affected by the same underlying condition. There are also moments where the connection runs through treatment – skin medications that impact the liver, or liver medications that cause skin reactions.

Vascular clues

One of the most distinctive skin signs of liver dysfunction is the spider angioma. These are small vascular lesions with a central red dot and fine capillaries that radiate outwards like the legs of a spider. You will often see them scattered across the chest, face, and upper limbs. They are not always pathological. Children get them all the time – my own son has one on his cheek as I write this. Yet, when they appear in clusters in an adult, especially on the upper torso, they are a classic sign of liver disease – particularly cirrhosis.

Palmar erythema is another vascular sign. It causes mottled redness over the palms that may blanch under pressure. It is not specific to liver disease alone, as it can appear in pregnancy or rheumatoid arthritis, but when seen alongside other features, it becomes a valuable clue. In more advanced cases, engorged abdominal veins can become visible and prominent. This is called caput medusae, named after the mythical Greek figure with snakes for hair.

Itch without a rash

One of the most troubling symptoms of chronic liver disease, especially in conditions that block or slow the flow of bile, is itching. This itch is deeper, more intense, and often feels like it comes from inside the body. It tends to be worst at night and is commonly felt on the palms of the hands and soles of the feet. There's usually no visible rash, but the discomfort can be overwhelming.

-)-)-)-
·:·:·:·

Hair, hormones, and fragility

Liver dysfunction affects the metabolism of sex hormones, and this often shows up in body hair. In men with advanced cirrhosis, you may see a more feminized pattern of hair distribution, with thinning of the beard and on the chest, and loss of pubic and armpit hair. This is due to the imbalance between oestrogen and androgens that develops as liver function declines.

The skin itself becomes fragile and easy to bruise. That is partly down to impaired production of clotting factors, but also reflects poor collagen support and weakened blood vessels.

A metabolic link between liver and skin

Porphyria cutanea tarda (PCT) is a condition where a fault in liver metabolism shows up most clearly on the skin. It causes fragile skin, blistering, and scarring, particularly on the backs of the hands and forearms, along with excess hair growth in those areas. Sunlight triggers the symptoms, but the root problem lies in how the liver processes porphyrins – natural chemicals involved in making haem (the pigment in blood). When porphyrins build up, they make the skin highly sensitive to light. PCT is closely linked with hepatitis C, alcohol use, and iron overload. Treatment often involves avoiding sun exposure and reducing excess iron through regular blood removal (phlebotomy).

• JAUNDICE •

The most well-known sign of liver disease on the skin is jaundice. This yellowing of the skin is caused by a build-up of bilirubin, the yellow pigment formed when red blood cells break down. When the liver is under strain, bilirubin accumulates in the blood and leads to the yellowing of the skin and eyes.

WHAT CAN SWOLLEN RED LEGS SAY ABOUT MY GENERAL HEALTH?

A hot, painful, rapidly spreading red patch on one leg, with sharp edges, tenderness, fever, or shivers is often cellulitis – bacteria in the deeper layers of the skin. This needs prompt antibiotics and sometimes hospital treatment. The tricky part is that a lot of non-infectious problems can look similar. If both legs are involved, it is more likely to be a type of dermatitis related to a problem with the vein circulation – this is not an infection, but can often get red and hot from the inflammation. Swelling in both legs that comes on more gradually, especially around the ankles and shins, and leaves a dent if you press with a finger, can reflect fluid overload rather than a primary skin problem. Heart failure, kidney disease, liver disease, and an adverse reaction to some blood pressure tablets all show themselves in the lower legs first. New unilateral swelling, particularly with tightness, pain in the calf, and a feeling of heaviness, raises the possibility of a deep vein thrombosis; this is a blood clot issue, not a skin issue, and needs urgent assessment rather than creams. As a rule of thumb, red swollen legs with fever, severe pain, shortness of breath, or sudden asymmetry are an emergency; chronic, bilateral changes with brown staining, itch, and eczema signal a long-term circulation problem that deserves proper vein, heart, and metabolic assessment.

—

WHAT IS IMPETIGO?

Impetigo is a very contagious superficial skin infection, usually caused by *Staphylococcus aureus*, but sometimes *Streptococcus*, that sits in the top layers of the epidermis. It often starts at a minor break in the skin, especially around the nose and mouth where skin is a little sore from wiping, and then blossoms into small red spots that quickly become blisters or moist patches that ooze and dry into a classic golden-yellow honey-coloured crust. The good news is that in healthy people it is usually straightforward to clear with the right antibiotics – topical or oral, depending on extent. Other basic measures that help include gentle cleansing, using separate towels, hand hygiene, and keeping children off school or nursery until the lesions are drying and no new ones are appearing.

—

ARE NUTRITIONAL SUPPLEMENTS WORTHWHILE?

For most generally healthy people eating a reasonably varied diet, routine nutritional supplements are far less important than the marketing would suggest. For the skin in particular, they are often a very expensive way to buy relatively ordinary vitamins. Where supplements are clearly worthwhile is when there is a proven or very likely deficiency or increased need: vitamin D if you live in a low-sunlight climate or cover your skin; iron or B12 if blood tests show you are low in iron or you are vegan; folate in pregnancy; calcium and vitamin D in some people at risk of osteoporosis; and carefully prescribed supplements in coeliac disease, inflammatory bowel disease, or after bariatric surgery. In these settings, correcting the deficiency can improve energy, hair shedding, skin dryness, and nail fragility, but the key is that it is targeted and monitored. From a dermatologist's point of view, your money and effort almost always go further if you invest them in a healthy general diet that is beneficial to your skin and the rest of your body every day.

—

WHAT IS THE BEST WAY TO MANAGE
THE CHICKENPOX VIRUS?

With chickenpox, the main job on the skin is to keep the lesions as comfortable, intact, and clean as possible while the immune system clears the virus. The spots follow a fairly predictable path – red bumps, then fluid-filled blisters, then cloudy blisters, then scabs – and the less they are scratched, the less the risk of infection and scarring. Keep your child's nails short and clean, while daily lukewarm baths with a bland emollient or a teaspoon of bicarbonate of soda in the water can soothe the skin. Any spots near the eyes, rapidly spreading redness, high fever, breathing difficulties, extreme sleepiness or confusion, or chickenpox in a very young baby, pregnant person, or someone immunosuppressed should prompt urgent medical assessment rather than home management alone.

—

epilogue

By now, you have learned the basics of how the skin is built and how it works. You have been given tools to understand your own skin and, I hope, the skin of the people you love. You have also seen how skin diseases and common skin concerns do not just sit quietly on the surface. They can itch, crack, bleed, sting, and disturb a good night's sleep. But they can also erode confidence, narrow down what you feel able to wear, where you feel able to go, and even who you feel able to be.

One of the most important messages I want you to carry away from this book is that you do not have to "put up with" most skin problems. Redness, flushing, constant breakouts, scaling patches on the scalp, an overly-sensitive face that stings with everything you put on it – none of these are things you deserve simply because you have difficult skin or are getting older. They are usually signs that something has gone off track, and in the vast majority of cases, there is something we can do to make things better.

At the same time, I hope I have been honest that there are some conditions we cannot cure outright. For these, the goal is management rather than eradication, learning how to spot the early warning signs, how to calm flares, and how to build routines and environments that let your skin stay as stable and comfortable as possible. That can sound less glamorous than a miracle cure, but in real life it is far more realistic and powerful. A well-managed chronic condition often occupies less of your day, and less of your mind, than a poorly acknowledged one that is allowed to rumble on in the background.

You have also seen that appearance and health are not separate stories. There is evidence that people who are perceived to look younger than their chronological age tend, on average, to live longer and have lower rates of certain diseases, and that long-term sun protection reduces both skin cancer risk and the visible signs of photo-ageing. When you apply sunscreen, treat skin harmed by UV radiation, or tackle chronic inflammation, you are reducing cumulative damage to the largest organ in your body.

With advances in medicine, most of us are now expecting to live longer than our parents and grandparents. That is a wonderful prospect – but it also means we will be asking a great deal more of our skin. This organ has to protect us, regulate us, sense the world for us, and keep doing it all, year after year, decade after decade. Caring for it should not be seen as a matter of vanity; it is part of how we look after our long-term health and, in a very real sense, our longevity.

There is a psychological story running alongside every skin story. Living with a visible skin condition can be exhausting. It can affect intimacy, work, friendships, and how you move through the

-)-)-)-

world. It is entirely reasonable to feel upset, angry, self-conscious, or simply tired of thinking about it. Asking for emotional support – from family, friends, or clinicians – can be a very useful part of optimizing skincare. Stress, low mood, poor sleep, and anxiety can make many skin conditions worse, and in turn, the skin can reinforce those same emotions. Breaking that loop often needs both medical treatment for the skin and compassionate attention for your mental health.

As you read this, we are at an interesting moment in dermatology. In the pages of this book I have had to be fairly general – explaining patterns, principles, and common pathways, rather than prescribing for you as an individual. But outside these pages, the field is moving rapidly towards more personalized care. We now understand more about genetics, the immune system, the microbiome, and how lifestyle, hormones, and the environment interact. Technology is allowing us to see and measure the skin in new ways, and to tailor treatments more precisely. Over the coming years, I expect we will move further away from one-size-fits-all routines and closer to plans that really do fit the person, their biology, and their life.

Even so, the foundations are unlikely to change: protect your skin from unnecessary damage; pay attention to changes and act on them early; and be cautious of quick fixes and miracle claims. Remember that making your skin as healthy as it can be almost always has benefits for the rest of you: less chronic inflammation, better sleep, more comfortable movement through the world, and often a quieter mind.

It has been my absolute privilege and ongoing fascination to spend my days thinking about skin – marvelling at the work it does when it is well, and helping people when it goes wrong. In clinic, I see every week how even small improvements can change how someone feels about themselves and their life. I hope that, in some small way, this book has helped you see your own skin differently and marvel at the wonderful job that it does – even if it sometimes misbehaves!

Share what you have learned. Ask questions. Be curious about your own skin and kind to other people in theirs.

All my very best to you and your skin,

Dr Emma Craythorne

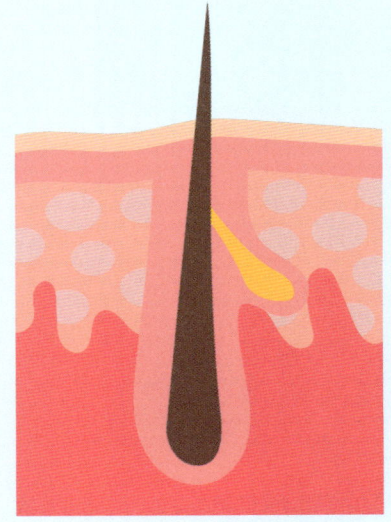

sources

Chapter 1 / Skin deep
Arda O, Göksügür N, Tüzün Y. Basic histological structure and functions of facial skin. Clin Dermatol. 2014 Jan-Feb;32(1):3–13. doi: 10.1016/*J. clindermatol.2013.05.021*. PMID: 24314373. / Polito MP, Marini G, Palamenghi M, Enzo E. Decoding the Human Epidermal Complexity at Single-Cell Resolution. *Int J Mol Sci.* 2023 May 10;24(10):8544. doi: 10.3390/ijms24108544. PMID: 37239891; PMCID: PMC10217868. / Wong R, Geyer S, Weninger W, Guimberteau JC, Wong JK. The dynamic anatomy and patterning of skin. *Exp Dermatol.* 2016 Feb;25(2):92–8. doi: 10.1111/exd.12832. Epub 2015 Oct 13. PMID: 26284579. / Kanitakis J. Anatomy, histology and immunohistochemistry of normal human skin. *Eur J Dermatol.* 2002 Jul-Aug;12(4):390–9; quiz 400–1. PMID: 12095893. / Chambers ES, Vukmanovic-Stejic M. Skin barrier immunity and ageing. *Immunology.* 2020 Jun;160(2):116–125. doi: 10.1111/imm.13152. Epub 2019 Dec 4. PMID: 31709535; PMCID: PMC7218662. / Kabashima K, Honda T, Ginhoux F, Egawa G. The immunological anatomy of the skin. *Nat Rev Immunol.* 2019 Jan;19(1):19–30. doi: 10.1038/s41577-018-0084-5. PMID: 30429578. / Zhang C, Merana GR, Harris-Tryon T, Scharschmidt TC. Skin immunity: dissecting the complex biology of our body's outer barrier. *Mucosal Immunol.* 2022 Apr;15(4):551–561. doi: 10.1038/s41385-022-00505-y. Epub 2022 Mar 31. PMID: 35361906. / Ali SM, Yosipovitch G. Skin pH: from basic science to basic skin care. *Acta Derm Venereol.* 2013 May;93(3):261–7. doi: 10.2340/00015555-1531. PMID: 23322028. / Rippke F, Schreiner V, Schwanitz HJ. The acidic milieu of the horny layer: new findings on the physiology and pathophysiology of skin pH. *Am J Clin Dermatol.* 2002;3(4):261–72. doi: 10.2165/00128071-200203040-00004. PMID: 12010071. / Proksch E. pH in nature, humans and skin. *J Dermatol.* 2018 Sep;45(9):1044–1052. doi: 10.1111/1346-8138.14489. Epub 2018 Jun 4. PMID: 29863755. / Whiting C, Abdel Azim S, Friedman A. The Skin Microbiome and its Significance for Dermatologists. *Am J Clin Dermatol.* 2024 Mar;25(2):169–177. doi: 10.1007/s40257-023-00842-z. Epub 2024 Jan 22. PMID: 38252188. / Flowers L, Grice EA. The Skin Microbiota: Balancing Risk and Reward. *Cell Host & Microbe.* 2020 Aug 12;28(2):190–200. doi: 10.1016/j.chom.2020.06.017. PMID: 32791112; PMCID: PMC7444652.

Chapter 2 / The great communicator
Handler A, Ginty DD. The mechanosensory neurons of touch and their mechanisms of activation. *Nat Rev Neurosci.* 2021 Sep;22(9):521–537. doi: 10.1038/s41583-021-00489-x. Epub 2021 Jul 26. PMID: 34312536; PMCID: PMC8485761. / Zimmerman A, Bai L, Ginty DD. The gentle touch receptors of mammalian skin. *Science.* 2014 Nov 21;346(6212):950–4. doi: 10.1126/science.1254229. PMID: 25414303; PMCID: PMC4450345. / Abraira VE, Ginty DD. The sensory neurons of touch. *Neuron.* 2013 Aug 21;79(4):618–39. doi: 10.1016/j.neuron.2013.07.051. PMID: 23972592; PMCID: PMC3811145. / Boulais N, Misery L. The epidermis: a sensory tissue. *Eur J Dermatol.* 2008 Mar-Apr;18(2):119–27. doi: 10.1684/ejd.2008.0348. PMID: 18424369. / Dubin AE, Patapoutian A. Nociceptors: the sensors of the pain pathway. *J Clin Invest.* 2010 Nov;120(11):3760–72. doi: 10.1172/JCI42843. Epub 2010 Nov 1. PMID: 21041958; PMCID: PMC2964977. / Luo J, Feng J, Liu S, Walters ET, Hu H. Molecular and cellular mechanisms that initiate pain and itch. *Cell Mol Life Sci.* 2015 Sep;72(17):3201–23. doi: 10.1007/s00018-015-1904-4. Epub 2015 Apr 18. Erratum in: *Cell Mol Life Sci.* 2015 Sep;72(18):3587–8. doi: 10.1007/s00018-015-1979-y. PMID: 25894692; PMCID: PMC4534341. / Calvo M, Dawes JM, Bennett DL. The role of the immune system in the generation of neuropathic pain. *Lancet Neurol.* 2012 Jul;11(7):629–42. doi: 10.1016/S1474-4422(12)70134-5. PMID: 22710756. / Misery L, Pierre O, Le Gall-Ianotto C, Lebonvallet N, Chernyshov PV, Le Garrec R, Talagas M. Basic mechanisms of itch. *J Allergy Clin Immunol.* 2023 Jul;152(1):11–23. doi: 10.1016/j.jaci.2023.05.004. Epub 2023 May 16. PMID: 37201903. / Sutaria N, Adawi W, Goldberg R, Roh YS, Choi J, Kwatra SG. Itch: Pathogenesis and treatment. *J Am Acad Dermatol.* 2022 Jan;86(1):17–34. doi: 10.1016/j.jaad.2021.07.078. Epub 2021 Oct 12. PMID: 34648873. / Coscarella G, Edwards E, Yosipovitch G. Basic mechanisms of itch and advances in clinical management. *Ann Allergy Asthma Immunol.* 2025 Oct 2:S1081-1206(25)01200-1. doi: 10.1016/j.anai.2025.09.014. Epub ahead of print. PMID: 41046125. / Pascalau R, Kuruvilla R. A Hairy End to a Chilling Event. *Cell.* 2020 Aug 6;182(3):539–541. doi: 10.1016/j.cell.2020.07.004. PMID: 32763185. / Lloyd DM, McGlone FP, Yosipovitch G. Somatosensory pleasure circuit: from skin to brain and back. *Exp Dermatol.* 2015 May;24(5):321–4. doi: 10.1111/exd.12639. Epub 2015 Mar 9. PMID: 25607755. / Crowley JS, Silverstein ML, Reghunathan M, Gosman AA. Glabellar Botulinum Toxin Injection Improves Depression Scores: A Systematic Review and Meta-Analysis. *Plast Reconstr Surg.* 2022 Jul 1;150(1):211e–220e. doi: 10.1097/PRS.0000000000009240. Epub 2022 May 20. PMID: 35588104. / Alam M, Barrett KC, Hodapp RM, Arndt KA. Botulinum toxin and the facial feedback hypothesis: can looking better make you feel happier? *J Am Acad Dermatol.* 2008 Jun;58(6):1061–72. doi: 10.1016/j.jaad.2007.10.649. PMID: 18485989. / Efthimiou TN, Baker J, Clarke A, Elsenaar A, Mehu M, Korb S. Zygomaticus activation through facial neuromuscular electrical stimulation (fNMES) induces happiness perception in ambiguous facial expressions and affects neural correlates of face processing. *Soc Cogn Affect Neurosci.* 2024 Feb 15;19(1):nsae013. doi: 10.1093/scan/nsae013. PMID: 38334739; PMCID: PMC10873823.

Chapter 3 / Skin through the ages
Schachner L, Andriessen A, Benjamin L, Bree A, Lechman P, Pinera-Llano A, Kircik L, Hebert A. The Importance of Skincare for Neonates and Infants: An Algorithm. *J Drugs Dermatol.* 2021 Nov 1;20(11):1195–1205. doi: 10.36849/jdd.6219. PMID: 34784132. / Afsar FS. Skin care for preterm and term neonates. *Clin Exp Dermatol.* 2009 Dec;34(8):855–8. doi: 10.1111/j.1365-2230.2009.03424.x. Epub 2009 Jul 2. PMID: 19575734. / Blume-Peytavi U, Hauser M, Stamatas GN, Pathirana D, Garcia Bartels N. Skin care practices for newborns and infants: review

of the clinical evidence for best practices. *Pediatr Dermatol.* 2012 Jan–Feb;29(1):1–14. doi: 10.1111/j.1525-1470.2011.01594.x. Epub 2011 Oct 20. PMID: 22011065. / Snyder KAM, Voelckers AD. Newborn Skin: Part I. Common Rashes and Skin Changes. *Am Fam Physician.* 2024 Mar;109(3):212–216. PMID: 38574210. / Schachner LA, Andriessen A, Benjamin L, Gonzalez ME, Kwong P, Woolery-Lloyd H, Heath C. Racial and Ethnic Variations in Skin Barrier Properties and Cultural Practices in Skin of Color Newborns, Infants, and Children. *J Drugs Dermatol.* 2023 Jul 1;22(7):657–663. doi: 10.36849/JDD.7305. PMID: 37410048. / Schachner L, Alexis A, Andriessen A, Baldwin H, Cork M, Kirsner R, Woolery-Lloyd H. Supplement Individual Article: The Importance of a Healthy Skin Barrier From the Cradle to the Grave Using Ceramide-Containing Cleansers and Moisturizers: A Review and Consensus. *J Drugs Dermatol.* 2023 Feb 1;22(2): SF344607s3-SF344607s14. PMID: 36745380. / LeFevre NM, Braudis K, Feigenbaum LS. Seborrheic Dermatitis: Diagnosis and Treatment. *Am Fam Physician.* 2025 Aug;112(2):166–173. PMID: 40834373. / Kastarinen H, Oksanen T, Okokon EO, Kiviniemi VV, Airola K, Jyrkkä J, Oravilahti T, Rannanheimo PK, Verbeek JH. Topical anti-inflammatory agents for seborrhoeic dermatitis of the face or scalp. *Cochrane Database Syst Rev.* 2014 May 19;2014(5):CD009446. doi: 10.1002/14651858.CD009446.pub2. PMID: 24838779; PMCID: PMC6483543. / Turgeon EW. Adolescent Skin: How to Keep it Healthy. *Can Fam Physician.* 1986 Nov; 32:2427–33. PMID: 21267224; PMCID: PMC2327993. / Hall G, Phillips TJ. Estrogen and skin: the effects of estrogen, menopause, and hormone replacement therapy on the skin. *J Am Acad Dermatol.* 2005 Oct;53(4):555–68; quiz 569-72. doi: 10.1016/j.jaad.2004.08.039. PMID: 16198774. / Raine-Fenning NJ, Brincat MP, Muscat-Baron Y. Skin aging and menopause: implications for treatment. *Am J Clin Dermatol.* 2003;4(6):371–8. doi: 10.2165/00128071-200304060-00001. PMID: 12762829. / Shu YY, Maibach HI. Estrogen and skin: therapeutic options. *Am J Clin Dermatol.* 2011 Oct 1;12(5):297–311. doi: 10.2165/11589180-000000000-00000. PMID: 21714580. / Crandall CJ, Mehta JM, Manson JE. Management of Menopausal Symptoms: A Review. *JAMA.*2023;329(5):405–420. doi:10.1001/jama.2022.24140 / Geraghty LN, Pomeranz MK. Physiologic changes and dermatoses of pregnancy. *Int J Dermatol.* 2011 Jul;50(7):771–82. doi: 10.1111/j.1365-4632.2010.04869.x. PMID: 21699510. / Wyles SP, Maredia HS, Ansaf RB, Dweydari MR, Hurt RT, Bonnes SL, Khosla S, LeBrasseur NK, Draelos ZD, Davis MDP. SkinspanTM: A Healthy Longevity Framework for Skin Aging. *Mayo Clin Proc.* 2025 Nov;100(11):1976–1991. doi: 10.1016/j.mayocp.2025.07.027. Epub 2025 Oct 1. PMID: 41032001. / Zhang J, Yu H, Man MQ, Hu L. Aging in the dermis: Fibroblast senescence and its significance. *Aging Cell.* 2024 Feb;23(2):e14054. doi: 10.1111/acel.14054. Epub 2023 Dec 1. PMID: 38040661; PMCID: PMC10861215. / Lee H, Hong Y, Kim M. Structural and Functional Changes and Possible Molecular Mechanisms in Aged Skin. *Int J Mol Sci.* 2021 Nov 19;22(22):12489. doi: 10.3390/ijms222212489. PMID: 34830368; PMCID: PMC8624050. / Rittié L, Fisher GJ. Natural and sun-induced aging of human skin. *Cold Spring Harb Perspect Med.* 2015 Jan 5;5(1):a015370. doi: 10.1101/cshperspect.a015370. PMID: 25561721; PMCID: PMC4292080. / Wahhab R, Sanders M, Kokikian N, Ni C, Vandiver AR. Clinical consequences of age-related skin barrier dysfunction–Part I: Structural, molecular, and physiologic changes with cutaneous aging. *J Am Acad Dermatol.* 2025 Oct;93(4):905–914. doi: 10.1016/j.jaad.2024.08.044. Epub 2024 Aug 29. PMID: 39216821. / Misery L, Taïeb C, Schollhammer M, Bertolus S, Coulibaly E, Feton-Danou N, Michel L, Seznec JC, Versapuech J, Joly P, Corgibet F, Ezzedine K, Richard MA. Psychological Consequences of the Most Common Dermatoses: Data from the Objectifs Peau Study. *Acta Derm Venereol.* 2020 Jun 11;100(13):adv00175. doi: 10.2340/00015555-3531. PMID: 32449783; PMCID: PMC9175061. / Christensen RE, Jafferany M. Psychiatric and psychologic aspects of chronic skin diseases. *Clin Dermatol.* 2023 Jan–Feb;41(1):75–81. doi: 10.1016/j.clindermatol.2023.03.006. Epub 2023 Mar 5. PMID: 36878453. / Dalgard FJ, Gieler U, Tomas-Aragones L, Lien L, Poot F, Jemec GBE, Misery L, Szabo C, Linder D, Sampogna F, Evers AWM, Halvorsen JA, Balieva F, Szepietowski J, Romanov D, Marron SE, Altunay IK, Finlay AY, Salek SS, Kupfer J. The psychological burden of skin diseases: a cross-sectional multicenter study among dermatological out-patients in 13 European countries. *J Invest Dermatol.* 2015 Apr;135(4):984–991. doi: 10.1038/jid.2014.530. Epub 2014 Dec 18. PMID: 25521458; PMCID: PMC4378256. / Stuhlmann CFZ, Traxler J, Paucke V, da Silva Burger N, Sommer R. Predictors and mechanisms of self-stigma in five chronic skin diseases: A systematic review. *J Eur Acad Dermatol Venereol.* 2025 Mar;39(3):622–630. doi: 10.1111/jdv.20314. Epub 2024 Sep 9. PMID: 39247975; PMCID: PMC11851255.

Chapter 4 / Conditions uncovered

Acne Vulgaris Author: Andrea L. Zaenglein, M.D N Engl J Med 2018;379:1343-1352DOI: 10.1056/NEJMcp1702493 / Eichenfield DZ, Sprague J, Eichenfield LF. Management of Acne Vulgaris: A Review. *JAMA.* 2021;326(20):2055–2067. doi:10.1001/jama.2021.17633 / Guidelines of care for the management of acne vulgaris. Reynolds, Rachel V. et al. *Journal of the American Academy of Dermatology,* Volume 90, Issue 5, 1006.e1 - 1006.e30 / Ständer S. Atopic Dermatitis. *N Engl J Med.* 2021 Mar 25;384(12):1136-1143. doi: 10.1056/NEJMra2023911. PMID: 33761208. / Weidinger S, Beck LA, Bieber T, Kabashima K, Irvine AD. Atopic dermatitis. *Nat Rev Dis Primers.* 2018 Jun 21;4(1):1. doi: 10.1038/s41572-018-0001-z. PMID: 29930342. / Grayson M. Psoriasis. *Nature.* 2012 Dec 20;492(7429):S49. doi: 10.1038/492S49a. PMID: 23254969. / Rosacea Esther J. van Zuuren, M.D. N Engl J Med 2017;377:1754-1764DOI: 10.1056/NEJMcp1506630 / van Zuuren EJ, Arents BWM, van der Linden MMD, Vermeulen S, Fedorowicz Z, Tan J. Rosacea: New Concepts in Classification and Treatment. *Am J Clin Dermatol.* 2021 Jul;22(4):457–465. doi: 10.1007/s40257-021-00595-7. Epub 2021 Mar 23. PMID: 33759078; PMCID: PMC8200341. / Ali L, Al Niaimi F. Pathogenesis of Melasma Explained. *Int J Dermatol.* 2025 Jul;64(7):1201–1212. doi: 10.1111/ijd.17718. Epub 2025 Feb 28. PMID: 40022484; PMCID: PMC12207721. / Passeron T, Picardo M. Melasma, a photoaging disorder. *Pigment Cell Melanoma Res.* 2018 Jul;31(4):461–465. doi: 10.1111/pcmr.12684. Epub 2018 Jan 12. PMID: 29285880. / Griffiths CEM, Armstrong AW, Gudjonsson JE, Barker JNWN. Psoriasis. *Lancet.* 2021 Apr 3;397(10281):1301–1315. doi: 10.1016/S0140-6736(20)32549-6. PMID: 33812489. / Raharja A, Mahil SK, Barker JN. Psoriasis: a brief overview. *Clin Med (Lond).* 2021 May;21(3):170–173. doi: 10.7861/clinmed.2021-0257. PMID: 34001566; PMCID: PMC8140694. / Upadhya S, Andrade MJ, Shukla V, Rao R, Satyamoorthy K. Genetic and immune dysregulation in vitiligo: Insights into autoimmune mechanisms and disease pathogenesis. *Autoimmun Rev.* 2025 Jul 31;24(8):103841. doi: 10.1016/j.autrev.2025.103841. Epub 2025 Jun 3. PMID: 40466982. / Bergqvist C, Ezzedine K. Vitiligo: A focus on pathogenesis and its therapeutic implications. *J Dermatol.* 2021 Mar;48(3):252-270. doi: 10.1111/1346-8138.15743. Epub 2021 Jan 6. PMID: 33404102. / Abdel-Malek ZA, Jordan C, Ho T, Upadhyay PR, Fleischer A, Hamzavi I. The enigma and challenges of vitiligo pathophysiology and treatment. *Pigment Cell Melanoma Res.* 2020 Nov; 33(6):778-787. doi: 10.1111/pcmr.12878. Epub 2020 Apr 12. PMID: 32198977. / Alessandrini A, Bruni F, Piraccini BM, Starace M. Common causes of hair loss – clinical manifestations, trichoscopy and therapy. *J Eur Acad Dermatol Venereol.* 2021 Mar;35(3):629-640. doi: 10.1111/jdv.17079. Epub 2021 Jan 8. PMID: 33290611. / Starace M, Orlando G, Alessandrini A, Piraccini BM. Female Androgenetic Alopecia: An Update on Diagnosis and Management. *Am J Clin Dermatol.* 2020 Feb;21(1):69-84. doi: 10.1007/s40257-019-00479-x. PMID: 31677111. / Kinoshita-Ise M, Fukuyama M, Ohyama M. Recent Advances in Understanding of the Etiopathogenesis, Diagnosis, and Management of Hair Loss Diseases. *J Clin Med.* 2023 May 3;12(9):3259. doi: 10.3390/jcm12093259. PMID: 37176700; PMCID: PMC10179687. / Zuberbier T, Abdul Latiff AH, Abuzakouk M, Aquilina S, Asero R, Baker D, Ballmer-Weber B, Bangert C, Ben-Shoshan M, Bernstein JA, Bindslev-Jensen C, Brockow K, Brzoza Z, Chong Neto HJ, Church MK, Criado PR, Danilycheva IV, Dressler C, Ensina LF, Fonacier L, Gaskins M, Gáspár K, Gelincik A, Giménez-Arnau A, Godse K, Gonçalo M, Grattan C, Grosber M, Hamelmann E, Hébert J, Hide M, Kaplan A, Kapp A, Kessel A, Kocatürk E, Kulthanan K, Larenas-Linnemann D, Lauerma A, Leslie TA, Magerl M, Makris M, Meshkova RY, Metz M, Micallef D, Mortz CG, Nast A, Oude-Elberink H, Pawankar R, Pigatto PD, Ratti Sisa H, Rojo Gutiérrez MI, Saini SS, Schmid-Grendelmeier P, Sekerel BE, Siebenhaar F, Siiskonen H, Soria A, Staubach-Renz P, Stingeni L, Sussman G, Szegedi A, Thomsen SF, Vadasz Z, Vestergaard C, Wedi B, Zhao Z, Maurer M. The international EAACI/GA²LEN/EuroGuiDerm/APAAACI guideline for the definition, classification, diagnosis, and management of urticaria. *Allergy.* 2022 Mar;77(3):734-766. doi: 10.1111/all.15090. Epub 2021 Oct 20. PMID: 34536239. / Radonjic-Hoesli S, Hofmeier KS, Micaletto S, Schmid-Grendelmeier P, Bircher A, Simon D. Urticaria and Angioedema: an Update on Classification and Pathogenesis. *Clin Rev Allergy Immunol.* 2018 Feb;54(1):88-101. doi: 10.1007/s12016-017-8628-1. PMID: 28748365. / Ali FR, Craythorne EE. Management of skin cancer in the elderly. *J Geriatr Oncol.* 2016 May;7(3):219-20. doi: 10.1016/j.jgo.2016.03.007. Epub 2016 Apr 19. PMID: 27103311. / Perez M, Abisaad JA, Rojas KD, Marchetti MA, Jaimes N. Skin cancer: Primary, secondary, and tertiary prevention. Part I. *J Am Acad Dermatol.* 2022 Aug;87(2):255-268. doi: 10.1016/j.jaad.2021.12.066. Epub 2022 Feb 14. PMID: 35176397. / Wehner MR. Keratinocyte Carcinoma: A Review. *JAMA.* Published online October 30, 2025. doi:10.1001/jama.2025.18749 / Gunaydin SD, Arikan-Akdagli S, Akova M. Fungal infections of the skin and soft tissue. *Curr Opin Infect Dis.* 2020 Apr;33(2):130-136. doi: 10.1097/QCO.0000000000000630. PMID: 31990815. / Zuber TJ, Baddam K. Superficial fungal infection of the skin. Where and how it appears help determine therapy. *Postgrad Med.* 2001 Jan;109(1):117-20, 123-6, 131-2. doi: 10.3810/pgm.2001.01.830. PMID: 11198246. / Squamous-Cell Carcinoma of the Skin. Ashley Wysong., *N Engl J Med* 2023;388:2262-2273 DOI: 10.1056/NEJMra2206348 / Verhaegen PD, van Zuijlen PP, Pennings NM, van Marle J, Niessen FB, van der Horst CM, Middelkoop E. Differences in collagen architecture between keloid, hypertrophic scar, normotrophic scar, and normal skin: An objective histopathological analysis. *Wound Repair Regen.* 2009 Sep-Oct;17(5):649-56. doi: 10.1111/j.1524-475X.2009.00533.x. PMID: 19769718. / Ogawa R. Keloid and Hypertrophic Scars Are the Result of Chronic Inflammation in the Reticular Dermis. *Int J Mol Sci.* 2017 Mar 10;18(3):606. doi: 10.3390/ijms18030606. PMID: 28287424; PMCID: PMC5372622. / Jeschke MG, Wood FM, Middelkoop E, Bayat A, Teot L, Ogawa R, Gauglitz GG. Scars. *Nat Rev Dis Primers.* 2023 Nov 16;9(1):64. doi: 10.1038/s41572-023-00474-x. PMID: 37973792. / Lee EH, Nehal KS, Disa JJ. Benign and premalignant skin lesions. *Plast Reconstr Surg.* 2010 May;125(5):188e-198e. doi: 10.1097/PRS.0b013e3181d6e89a. PMID: 20440130. / Higgins JC, Maher MH, Douglas MS. Diagnosing Common Benign Skin Tumors. *Am Fam Physician.* 2015 Oct 1;92(7):601-7. PMID: 26447443. / Yurchenko AA, Rajabi F, Braz-Petta T, Fassihi H, Lehmann A, Nishigori C, Wang J, Padioleau I, Gunbin K, Panunzi L, Morice-Picard F, Laplante J, Robert C, Kannouche PL, Menck CFM, Sarasin A, Nikolaev SI. Genomic mutation landscape of skin cancers from DNA repair-deficient xeroderma pigmentosum patients. *Nat Commun.* 2023 May 4;14(1):2561. doi: 10.1038/s41467-023-38311-0. PMID: 37142601; PMCID: PMC10160032. / Fassihi H. Spotlight on 'xeroderma pigmentosum'. *Photochem Photobiol Sci.* 2013 Jan;12(1):78-84. doi: 10.1039/c2pp25267h. Erratum in: Photochem Photobiol Sci. 2014 Sep;13(9):1359. PMID: 23132518.

Chapter 5 / A skin loving lifestyle

Schuch AP, Moreno NC, Schuch NJ, Menck CFM, Garcia CCM. Sunlight damage to cellular DNA: Focus on oxidatively generated lesions. *Free Radic Biol Med.* 2017 Jun;107:110-124. doi: 10.1016/j.freeradbiomed.2017.01.029. Epub 2017 Jan 18. PMID: 28109890. / Marrot L, Meunier JR. Skin DNA photodamage and its biological consequences. *J Am Acad Dermatol.* 2008 May;58(5 Suppl 2):S139-48. doi: 10.1016/j.jaad.2007.12.007. PMID: 18410800. / D'Orazio J, Jarrett S, Amaro-Ortiz A, Scott T. UV radiation and the skin. *Int J Mol Sci.* 2013 Jun 7;14(6):12222-48. doi: 10.3390/ijms140612222. PMID: 23749111; PMCID: PMC3709783. / Liebel F, Kaur S, Ruvolo E, Kollias N, Southall MD. Irradiation of skin with visible light induces reactive oxygen species and matrix-degrading enzymes. *J Invest Dermatol.* 2012 Jul;132(7):1901-7. doi: 10.1038/jid.2011.476. Epub 2012 Feb 9. PMID: 22318388. / Puri P, Nandar SK, Kathuria S, Ramesh V. Effects of air pollution on the skin: A review. *Indian J Dermatol Venereol Leprol.* 2017 Jul-Aug;83(4):415-423. doi: 10.4103/0378-6323.199579. PMID: 28195077. / Ortiz A, Grando SA. Smoking and the skin. *Int J Dermatol.* 2012 Mar;51(3):250-62. doi: 10.1111/j.1365-4632.2011.05205.x. PMID: 22348557. / Morita A. Tobacco smoke causes premature skin aging. *J Dermatol Sci.* 2007 Dec;48(3):169-75. doi: 10.1016/j.jdermsci.2007.06.015. Epub 2007 Oct 24. PMID: 17951030. / Araviiskaia E, Berardesca E, Bieber T, Gontijo G, Sanchez Viera M, Marrot L, Chuberre B, Dreno B. The impact of airborne pollution on skin. *J Eur Acad Dermatol Venereol.* 2019 Aug;33(8):1496-1505. doi: 10.1111/jdv.15583. Epub 2019 Apr 26. PMID: 30897234; PMCID: PMC6766865. / Damevska K, Boev B, Mirakovski D, Petrov A, Darlenski R, Simeonovski V. How to prevent skin damage from air pollution. Part 1: Exposure assessment. *Dermatol Ther.* 2020 Jan;33(1):e13171. doi: 10.1111/dth.13171. Epub 2019 Dec 4. PMID: 31750979. / Abolhasani R, Araghi F, Tabary M, Aryannejad A, Mashinchi B, Robati RM. The impact of air pollution on skin and related disorders: A comprehensive review. *Dermatol Ther.* 2021 Mar;34(2):e14840. doi: 10.1111/dth.14840. Epub 2021 Feb 12. PMID: 33527709. / Bennett SL, Khachemoune A. Dispelling myths about sunscreen. *J Dermatolog Treat.* 2022 Mar;33(2):666-670. doi: 10.1080/09546634.2020.1789047. Epub 2020 Jul 7. PMID: 32633165. / Cao C, Xiao Z, Wu Y, Ge C. Diet and Skin Aging-From the Perspective of Food Nutrition. *Nutrients.* 2020 Mar 24;12(3):870. doi: 10.3390/nu12030870. PMID: 32213934; PMCID: PMC7146365. / Muzumdar S, Ferenczi K. Nutrition and youthful skin. *Clin Dermatol.* 2021 Sep-Oct;39(5):796-808. doi: 10.1016/j.clindermatol.2021.05.007. Epub 2021 May 13. PMID: 34785007. / Chen CT, Tung HH, Tung TH, Denq JC. Nutrition, Exercise, and Skin Integrity among Frail Older

Adults in Taiwan. *Adv Skin Wound Care*. 2017 Aug;30(8):364-371. doi: 10.1097/01.ASW.0000516309.92029.4e. PMID: 28727592. / Banks MD, Webster J, Bauer J, Dwyer K, Pelecanos A, MacDermott P, Nevin A, Coleman K, Campbell J, Hickling D, Byrnes A, Capra S. Effect of supplements/intensive nutrition on pressure ulcer healing: a multicentre, randomised controlled study. *J Wound Care*. 2023 May 2;32(5):292-300. doi: 10.12968/jowc.2023.32.5.292. PMID: 37094924. / Afzal UM, Ali FR. Sleep deprivation and the skin. *Clin Exp Dermatol*. 2023 Sep 19;48(10):1113-1116. doi: 10.1093/ced/llad196. PMID: 37288611. / Cameron S, Donnelly A, Broderick C, Arichi T, Bartsch U, Dazzan P, Elberling J, Godfrey E, Gringras P, Heathcote LC, Joseph D, Wood TC, Pariante C, Rubia K, Flohr C. Mind and skin: Exploring the links between inflammation, sleep disturbance and neurocognitive function in patients with atopic dermatitis. *Allergy*. 2024 Jan;79(1):26-36. doi: 10.1111/all.15818. Epub 2023 Jul 19. Erratum in: Allergy. 2024 Jun;79(6):1641. doi: 10.1111/all.16079. PMID: 37469218. / Oyetakin-White P, Suggs A, Koo B, Matsui MS, Yarosh D, Cooper KD, Baron ED. Does poor sleep quality affect skin ageing? *Clin Exp Dermatol*. 2015 Jan;40(1):17-22. doi: 10.1111/ced.12455. Epub 2014 Sep 30. PMID: 25266053. / Nishikori S, Yasuda J, Murata K, Takegaki J, Harada Y, Shirai Y, Fujita S. Resistance training rejuvenates aging skin by reducing circulating inflammatory factors and enhancing dermal extracellular matrices. *Sci Rep*. 2023 Jun 23;13(1):10214. doi: 10.1038/s41598-023-37207-9. PMID: 37353523; PMCID: PMC10290068. / Zhu WG, Thomas ACQ, Wilson GM, McGlory C, Hibbert JE, Flynn CG, Sayed RKA, Paez HG, Meinhold M, Jorgenson KW, You JS, Steinert ND, Lin KH, MacInnis MJ, Coon JJ, Phillips SM, Hornberger TA. Identification of a resistance-exercise-specific signalling pathway that drives skeletal muscle growth. *Nat Metab*. 2025 Jul;7(7):1404-1423. doi: 10.1038/s42255-025-01298-7. Epub 2025 May 15. PMID: 40374925. / Ertaş R, Türk M, Yücel MB, Muñoz M, Ertaş ŞK, Atasoy M, Maurer M. Eating Increases and Exercise Decreases Disease Activity in Patients With Symptomatic Dermographism. *J Allergy Clin Immunol Pract*. 2023 Mar;11(3):932-940. doi: 10.1016/j.jaip.2022.11.041. Epub 2022 Dec 16. PMID: 36535522. / Zhang H, Wang M, Zhao X, Wang Y, Chen X, Su J. Role of stress in skin diseases: A neuroendocrine-immune interaction view. *Brain Behav Immun*. 2024 Feb;116:286-302. doi: 10.1016/j.bbi.2023.12.005. Epub 2023 Dec 20. PMID: 38128623. / Hunter HJ, Momen SE, Kleyn CE. The impact of psychosocial stress on healthy skin. *Clin Exp Dermatol*. 2015 Jul;40(5):540-6. doi: 10.1111/ced.12582. Epub 2015 Mar 25. PMID: 25808947. / Kubrak A, Zimny-Zając A, Makuch S, Jankowska-Polańska B, Tański W, Szepietowski JC, Agrawal S. Psychosocial Factors, Stress, and Well-Being: Associations with Common Dermatological Manifestations in a Large Polish Cross-Sectional Analysis. *J Clin Med*. 2025 Jun 3;14(11):3943. doi: 10.3390/jcm14113943. PMID: 40507703; PMCID: PMC12156280. / Assaf S, Kelly O. Nutritional Dermatology: Optimizing Dietary Choices for Skin Health. *Nutrients*. 2024 Dec 27;17(1):60. doi: 10.3390/nu17010060. PMID: 39796494; PMCID: PMC11723311.

Chapter 6 / Guarding the barrier

Trepanowski N, Grant-Kels JM. Social media dermatologic advice: Dermatology without dermatologists. *JAAD Int*. 2023 May 26;12:101-102. doi: 10.1016/j.jdin.2023.05.004. PMID: 37404249; PMCID: PMC10315776. / Mays D, Friedman A, Kennedy J, Yiannias J, Morgan J. Non-adherence to Labeling Standards Can Misrepresent Safety of Ingredients in Cosmetic Cleansers. *J Drugs Dermatol*. 2023 Jan 1;22(1):98-100. PMID: 36607752. / Landau M, Landau SB. Hacking the International Nomenclature of Cosmetic Ingredients List- How to Read Ingredients in Cosmetic Products and What Is Important for a Dermatologist to Know? *Dermatol Clin*. 2024 Jan;42(1):7-11. doi: 10.1016/j.det.2023.06.006. Epub 2023 Jul 27. PMID: 37977687. / Urban K, Giesey R, Delost G. A Guide to Informed Skincare: The Meaning of Clean, Natural, Organic, Vegan, and Cruelty-Free. *J Drugs Dermatol*. 2022 Sep 1;21(9):1012-1013. doi: 10.36849/JDD.6795. PMID: 36074502. / Giordano-Labadie F. Cosmetic products: learning to read labels. *Eur J Dermatol*. 2012 Sep-Oct;22(5):591-5. doi: 10.1684/ejd.2012.1786. PMID: 22782011. / Baumann L. Understanding and treating various skin types: the Baumann Skin Type Indicator. *Dermatol Clin*. 2008 Jul;26(3):359-73, vi. doi: 10.1016/j.det.2008.03.007. PMID: 18555952. / Roberts WE. Skin type classification systems old and new. *Dermatol Clin*. 2009 Oct;27(4):529-33, viii. doi: 10.1016/j.det.2009.08.006. PMID: 19850202. / Fowler JF Jr. Routine skin care as prophylaxis and treatment. *Semin Cutan Med Surg*. 2013 Jun;32(2 Suppl 2):S15. doi: 10.12788/j.sder.0020. PMID: 24156152. / Flament F, Mercurio DG, Catalan E, Bouhadanna E, Delaunay C, Miranda DF, Passeron T. Impact on facial skin aging signs of a 1-year standardized photoprotection over a classical skin care routine in skin phototypes II-VI individuals: A prospective randomized trial. *J Eur Acad Dermatol Venereol*. 2023 Oct;37(10):2090-2097. doi: 10.1111/jdv.19230. Epub 2023 Jun 9. PMID: 37247191. / Lain E, Alexis AF, Andriessen A, Campos VB, Haus A, Kim J, Lupin M, McDonald C, Zhang CF. A Practical Algorithm for Integrating Skincare to Improve Patient Outcomes and Satisfaction With Energy-Based Dermatologic Procedures. *J Drugs Dermatol*. 2024 May 1;23(5):353-359. doi: 10.36849/JDD.8092. PMID: 38709701. / Alvarez GV, Kang BY, Richmond AM, Hoss E, Sulewski R, Minkis K, Rozenberg SS, Antonovich D, Boucher A, Bernstein EF, Bertucci V, Chapas AM, Cohen JL, Council ML, Dover JS, Geronemus R, Given KML, Goldbach HS, Goldman MP, Hooper D, Kaufman J, Munavalli G, Pacheco TR, Rossi AM, Wilson S, Alam M. Skincare ingredients recommended by cosmetic dermatologists: A Delphi consensus study. *J Am Acad Dermatol*. 2025 Apr 14:S0190-9622(25)00612-7. doi: 10.1016/j.jaad.2025.04.021. Epub ahead of print. PMID: 40233838. / Liu H, Yu H, Xia J, Liu L, Liu GJ, Sang H, Peinemann F. Topical azelaic acid, salicylic acid, nicotinamide, sulphur, zinc and fruit acid (alpha-hydroxy acid) for acne. *Cochrane Database Syst Rev*. 2020 May 1;5(5):CD011368. doi: 10.1002/14651858.CD011368.pub2. PMID: 32356169; PMCID: PMC7193765. / de Troya-Martín M, Rodríguez-Martínez A, Rivas-Ruiz F, Subert A, Arellano-Mendoza MI, Calzavara-Pinton P, de Gálvez MV, Gilaberte Y, Goh CL, Lim HW, Schalka S, Wolf P, González S. Personalized Photoprotection: Expert Consensus and Recommendations From a Delphi Study Among Dermatologists. *Photodermatol Photoimmunol Photomed*. 2025 Jan;41(1):e70001. doi: 10.1111/phpp.70001. PMID: 39868505; PMCID: PMC11771696. / Taylor S, Elbuluk N, Grimes P, Chien A, Hamzavi I, Alexis A, Gonzalez N, Weiss J, Kang S, Desai SR. Treatment recommendations for acne-associated hyperpigmentation: Results of the Delphi consensus process and a literature review. *J Am Acad Dermatol*. 2023 Aug;89(2):316-323. doi: 10.1016/j.jaad.2023.02.053. Epub 2023 Mar 15. PMID: 36924935. / Thiboutot D, Layton AM, Traore I, Gontijo G, Troielli P, Ju Q, Kurokawa I, Dreno B. International expert consensus recommendations for the use of dermocosmetics in acne. *J Eur Acad Dermatol Venereol*. 2025 May;39(5):952-966. doi: 10.1111/jdv.20145. Epub 2024 Jun 15. PMID: 38877766; PMCID: PMC12023719. / Proksch E, Brandner JM, Jensen JM. The skin: an indispensable barrier. *Exp Dermatol*. 2008 Dec;17(12):1063-72. doi: 10.1111/j.1600-0625.2008.00786.x. PMID: 19043850. / Sakuma TH, Maibach HI. Oily skin: an overview. *Skin*

Pharmacol Physiol. 2012;25(5):227-35. doi: 10.1159/000338978. Epub 2012 Jun 20. PMID: 22722766. / Jensen JM, Proksch E. The skin's barrier. G Ital Dermatol Venereol. 2009 Dec;144(6):689-700. PMID: 19907407. / Norlén L. Molecular Organization of the Skin Barrier. Acta Derm Venereol. 2023 Nov 21;103:adv13356. doi: 10.2340/actadv.v103.13356. PMID: 37987626; PMCID: PMC10680981. / Blaak J, Staib P. The Relation of pH and Skin Cleansing. Curr Probl Dermatol. 2018; 54:132-142. doi: 10.1159/000489527. Epub 2018 Aug 21. PMID: 30130782. / Draelos ZD. The science behind skin care: Cleansers. J Cosmet Dermatol. 2018 Feb;17(1):8-14. doi: 10.1111/jocd.12469. Epub 2017 Dec 12. PMID: 29231284. / Cowdell F, Steventon K. Skin cleansing practices for older people: a systematic review. Int J Older People Nurs. 2015 Mar;10(1):3-13. doi: 10.1111/opn.12041. Epub 2013 Oct 1. PMID: 24118822. / Rawlings AV, Harding CR. Moisturization and skin barrier function. Dermatol Ther. 2004;17 Suppl 1:43-8. doi: 10.1111/j.1396-0296.2004.04s1005.x. PMID: 14728698. / Lodén M. Role of topical emollients and moisturizers in the treatment of dry skin barrier disorders. Am J Clin Dermatol. 2003;4(11):771-88. doi: 10.2165/00128071-200304110-00005. PMID: 14572299.

Chapter 7 / The clinic

Mowad CM, Anderson B, Scheinman P, Pootongkam S, Nedorost S, Brod B. Allergic contact dermatitis: Patient management and education. J Am Acad Dermatol. 2016 Jun;74(6):1043-54. doi: 10.1016/j.jaad.2015.02.1144. PMID: 27185422. / Mowad CM, Anderson B, Scheinman P, Pootongkam S, Nedorost S, Brod B. Allergic contact dermatitis: Patient diagnosis and evaluation. J Am Acad Dermatol. 2016 Jun;74(6):1029-40. doi: 10.1016/j.jaad.2015.02.1139. PMID: 27185421. / Patel G, Saltoun C. Skin testing in allergy. Allergy Asthma Proc. 2019 Nov 1;40(6):366-368. doi: 10.2500/aap.2019.40.4248. PMID: 31690371. / Ahlstedt S. Understanding the usefulness of specific IgE blood tests in allergy. Clin Exp Allergy. 2002 Jan;32(1):11-6. doi: 10.1046/j.0022-0477.2001.01289.x. PMID: 12002727. / Mounsey AL, Reed SW. Diagnosing and treating hair loss. Am Fam Physician. 2009 Aug 15;80(4):356-62. PMID: 19678603. / Jackson AJ, Price VH. How to diagnose hair loss. Dermatol Clin. 2013 Jan;31(1):21-8. doi: 10.1016/j.det.2012.08.007. Epub 2012 Sep 29. PMID: 23159173. / Bacharier LB, Jackson DJ. Biologics in the treatment of asthma in children and adolescents. J Allergy Clin Immunol. 2023 Mar;151(3):581-589. doi: 10.1016/j.jaci.2023.01.002. Epub 2023 Jan 24. PMID: 36702649. / Griffiths CEM, Armstrong AW, Gudjonsson JE, Barker JNWN. Psoriasis. Lancet. 2021 Apr 3;397(10281):1301-1315. doi: 10.1016/S0140-6736(20)32549-6. PMID: 33812489. / Torres AE, Lyons AB, Hamzavi IH, Lim HW. Role of phototherapy in the era of biologics. J Am Acad Dermatol. 2021 Feb;84(2):479-485. doi: 10.1016/j.jaad.2020.04.095. Epub 2020 Apr 24. PMID: 32339702; PMCID: PMC7194984. / Lim HW, Silpa-archa N, Amadi U, Menter A, Van Voorhees AS, Lebwohl M. Phototherapy in dermatology: A call for action. J Am Acad Dermatol. 2015 Jun;72(6):1078-80. doi: 10.1016/j.jaad.2015.03.017. PMID: 25981004. / Ashique KT, Kaliyadan F, Jayasree P. Cryotherapy: Tips and Tricks. J Cutan Aesthet Surg. 2021 Apr-Jun;14(2):244-247. doi: 10.4103/JCAS.JCAS_141_20. PMID: 34566372; PMCID: PMC8423213. / Yang S, Kampp J. Common Dermatologic Procedures. Med Clin North Am. 2015 Nov;99(6):1305-21. doi: 10.1016/j.mcna.2015.07.004. Epub 2015 Sep 11. PMID: 26476254. / Bensimon RH. Chemical Peels. Facial Plast Surg Clin North Am. 2023 Nov;31(4):475-494. doi: 10.1016/j.fsc.2023.05.006. Epub 2023 Jul 26. PMID: 37806681. / Landau M, Bageorgeou F. Update on Chemical Peels. Dermatol Clin. 2024 Jan;42(1):13-20. doi: 10.1016/j.det.2023.06.005. Epub 2023 Jul 23. PMID: 37977680. / O'Connor AA, Lowe PM, Shumack S, Lim AC. Chemical peels: A review of current practice. Australas J Dermatol. 2018 Aug;59(3):171-181. doi: 10.1111/ajd.12715. Epub 2017 Oct 24. PMID: 29064096. / Martina E, Diotallevi F, Radi G, Campanati A, Offidani A. Therapeutic Use of Botulinum Neurotoxins in Dermatology: Systematic Review. Toxins (Basel). 2021 Feb 5;13(2):120. doi: 10.3390/toxins13020120. PMID: 33562846; PMCID: PMC7915854. / Goldberg DJ. Botulinum toxin injections have become a main-stay in the treatment of aging skin. J Cosmet Laser Ther. 2008 Jun;10(2):66. doi: 10.1080/14764170802200095. PMID: 18569257. / Costeloe A, Nguyen A, Maas C. Neuromodulators for Skin. Facial Plast Surg Clin North Am. 2023 Nov;31(4):511-519. doi: 10.1016/j.fsc.2023.06.002. Epub 2023 Jul 14. PMID: 37806684. / Kim JE, Sykes JM. Hyaluronic acid fillers: history and overview. Facial Plast Surg. 2011 Dec;27(6):523-8. doi: 10.1055/s-0031-1298785. Epub 2011 Dec 28. PMID: 22205525. / Chacon AH. Fillers in dermatology: from past to present. Cutis. 2015 Nov;96(5):E17-9. PMID: 26682563. / Lee JC, Lorenc ZP. Synthetic Fillers for Facial Rejuvenation. Clin Plast Surg. 2016 Jul;43(3):497-503. doi: 10.1016/j.cps.2016.03.002. PMID: 27363763. / Chen SX, Cheng J, Watchmaker J, Dover JS, Chung HJ. Review of Lasers and Energy-Based Devices for Skin Rejuvenation and Scar Treatment With Histologic Correlations. Dermatol Surg. 2022 Apr 1;48(4):441-448. doi: 10.1097/DSS.0000000000003397. PMID: 35165220. / Hanke CW. Lasers in dermatology. Indiana Med. 1990 Jun;83(6):394-402. PMID: 2112179. / Anderson RR, Parrish JA. Selective photothermolysis: precise microsurgery by selective absorption of pulsed radiation. Science. 1983 Apr 29;220(4596):524-7. doi: 10.1126/science.6836297. PMID: 6836297. / Patel AO, Chopra R, Avram M, Sakamoto FH, Kilmer S, Anderson RR, Ibrahimi OA. Updates on Lasers in Dermatology. Dermatol Clin. 2024 Jan;42(1):33-44. doi: 10.1016/j.det.2023.07.004. Epub 2023 Oct 17. PMID: 37977682. / Emer J. Platelet-Rich Plasma (PRP): Current Applications in Dermatology. Skin Therapy Lett. 2019 Sep;24(5):1-6. PMID: 31584784. / Vyas KS, Kaufman J, Munavalli GS, Robertson K, Behfar A, Wyles SP. Exosomes: the latest in regenerative aesthetics. Regen Med. 2023 Feb;18(2):181-194. doi: 10.2217/rme-2022-0134. Epub 2023 Jan 4. PMID: 36597716.

Chapter 8 / Common concerns

Hald M, Arendrup MC, Svejgaard EL, Lindskov R, Foged EK, Saunte DM; Danish Society of Dermatology. Evidence-based Danish guidelines for the treatment of Malassezia-related skin diseases. Acta Derm Venereol. 2015 Jan;95(1):12-9. doi: 10.2340/00015555-1825. PMID: 24556907. / Luelmo-Aguilar J, Santandreu MS. Folliculitis: recognition and management. Am J Clin Dermatol. 2004;5(5):301-10. doi: 10.2165/00128071-200405050-00003. PMID: 15554731. / Chi CC, Wang SH, Delamere FM, Wojnarowska F, Peters MC, Kanjirath PP. Interventions for prevention of herpes simplex labialis (cold sores on the lips). Cochrane Database Syst Rev. 2015 Aug 7;2015(8):CD010095. doi: 10.1002/14651858.CD010095.pub2. PMID: 26252373; PMCID: PMC6461191. / Gopinath D, Koe KH, Maharajan MK, Panda S. A

Comprehensive Overview of Epidemiology, Pathogenesis and the Management of Herpes Labialis. *Viruses.* 2023 Jan 13;15(1):225. doi: 10.3390/v15010225. PMID: 36680265; PMCID: PMC9867007. / Narayanan D, Rogge M. Cheilitis: A Diagnostic Algorithm and Review of Underlying Etiologies. *Dermatitis.* 2024 Sep-Oct;35(5):431-442. doi: 10.1089/derm.2023.0276. Epub 2024 Feb 29. PMID: 38422211. / Palaniappan V, Sadhasivamohan A, Sankarapandian J, Karthikeyan K. Miliaria crystallina. *Clin Exp Dermatol.* 2023 Apr 27;48(5):462-467. doi: 10.1093/ced/llad032. PMID: 36692206. / Feng E, Janniger CK. Miliaria. *Cutis.* 1995 Apr;55(4):213-6. PMID: 7796612. / Lashway SG, Worthen ADM, Abuasbeh JN, Harris RB, Farland LV, O'Rourke MK, Dennis LK. A meta-analysis of sunburn and basal cell carcinoma risk. *Cancer Epidemiol.* 2023 Aug;85:102379. doi: 10.1016/j.canep.2023.102379. Epub 2023 May 16. PMID: 37201363. / Groves GA. Sunburn and its prevention. *Australas J Dermatol.* 1980 Dec;21(3):115-41. doi: 10.1111/j.1440-0960.1980.tb00156.x. PMID: 7016109. / Weber I, Liao K, Dang T, Shah M, Wehner MR. Sunburn and Cutaneous Squamous Cell Carcinoma: A Meta-Analysis. *JAMA Dermatol.* 2025 Nov 1;161(11):1148-1156. doi: 10.1001/jamadermatol.2025.3473. PMID: 40991277; PMCID: PMC12461597. / Braun HA, Adler CH, Goodman M, Yeung H. Sunburn frequency and risk and protective factors: a cross-sectional survey. *Dermatol Online J.* 2021 Apr 15;27(4):13030/ qt6qn7k2gp. PMID: 33999575; PMCID: PMC8281353. / Kodali N, Patel VM, Schwartz RA. Keratosis pilaris: an update and approach to management. *Ital J Dermatol Venerol.* 2023 Jun;158(3):217-223. doi: 10.23736/S2784-8671.23.07594-1. Epub 2023 May 11. PMID: 37166753. / Wong PC, Wang MA, Ng TJ, Akbarialiabad H, Murrell DF. Keratosis pilaris treatment paradigms: assessing effectiveness across modalities. *Clin Exp Dermatol.* 2024 Sep 18;49(10):1105-1117. doi: 10.1093/ced/llae066. PMID: 38447098. / Ely JW, Rosenfeld S, Seabury Stone M. Diagnosis and management of tinea infections. *Am Fam Physician.* 2014 Nov 15;90(10):702-10. PMID: 25403034. / Thomas C, Coates SJ, Engelman D, Chosidow O, Chang AY. Ectoparasites: Scabies. *J Am Acad Dermatol.* 2020 Mar;82(3):533-548. doi: 10.1016/j.jaad.2019.05.109. Epub 2019 Jul 13. PMID: 31310840. / Bernigaud C, Fischer K, Chosidow O. The Management of Scabies in the 21st Century: Past, Advances and Potentials. *Acta Derm Venereol.* 2020 Apr 20;100(9):adv00112. doi: 10.2340/00015555-3468. PMID: 32207535; PMCID: PMC9128908. / Friedmann DP, Goldman MP. Dark circles: etiology and management options. *Clin Plast Surg.* 2015 Jan;42(1):33-50. doi: 10.1016/j.cps.2014.08.007. PMID: 25440739.

Chapter 9 / Systematic health
Hurwitz S. The skin and systemic disease in children. *Curr Probl Pediatr.* 1977 Jul;7(9):1-68. doi: 10.1016/s0045-9380(77)80008-x. PMID: 330114. / McDonald CJ. Cutaneous manifestations of systemic disease. *Postgrad Med.* 1981 May;69(5):132-5, 138-40, 143 passim. doi: 10.1080/00325481.1981.11715762. PMID: 7232242. / Robson KJ, Piette WW. Cutaneous manifestations of systemic diseases. *Med Clin North Am.* 1998 Nov;82(6):1359-79, vi-vii. doi: 10.1016/s0025-7125(05)70419-3. PMID: 9889752. / Sánchez-Pellicer P, Navarro-Moratalla L, Núñez-Delegido E, Ruzafa-Costas B, Agüera-Santos J, Navarro-López V. Acne, Microbiome, and Probiotics: The Gut-Skin Axis. *Microorganisms.* 2022 Jun 27;10(7):1303. doi: 10.3390/microorganisms10071303. PMID: 35889022; PMCID: PMC9318165. / Mahmud MR, Akter S, Tamanna SK, Mazumder L, Esti IZ, Banerjee S, Akter S, Hasan MR, Acharjee M, Hossain MS, Pirttilä AM. Impact of gut microbiome on skin health: gut-skin axis observed through the lenses of therapeutics and skin diseases. *Gut Microbes.* 2022 Jan-Dec;14(1):2096995. doi: 10.1080/19490976.2022.2096995. PMID: 35866234; PMCID: PMC9311318. / Sinha S, Lin G, Ferenczi K. The skin microbiome and the gut-skin axis. *Clin Dermatol.* 2021 Sep-Oct;39(5):829-839. doi: 10.1016/j.clindermatol.2021.08.021. Epub 2021 Sep 3. PMID: 34785010. / Salem I, Ramser A, Isham N, Ghannoum MA. The Gut Microbiome as a Major Regulator of the Gut-Skin Axis. *Front Microbiol.* 2018 Jul 10;9:1459. doi: 10.3389/fmicb.2018.01459. PMID: 30042740; PMCID: PMC6048199. / Lima AL, Illing T, Schliemann S, Elsner P. Cutaneous Manifestations of Diabetes Mellitus: A Review. *Am J Clin Dermatol.* 2017 Aug;18(4):541-553. doi: 10.1007/s40257-017-0275-z. PMID: 28374407. / Marchand L, Gaimard M, Luyton C. All about skin manifestations of insulin resistance and type 2 diabetes: acanthosis nigricans and acrochordons. *Postgrad Med J.* 2020 Apr;96(1134):237. doi: 10.1136/postgradmedj-2019-136834. Epub 2019 Oct 13. PMID: 31611265. / Karadag AS, Ozlu E, Lavery MJ. Cutaneous manifestations of diabetes mellitus and the metabolic syndrome. *Clin Dermatol.* 2018 Jan-Feb;36(1):89-93. doi: 10.1016/j.clindermatol.2017.09.015. Epub 2017 Sep 8. PMID: 29241758. / Lause M, Kamboj A, Fernandez Faith E. Dermatologic manifestations of endocrine disorders. *Transl Pediatr.* 2017 Oct;6(4):300-312. doi: 10.21037/tp.2017.09.08. PMID: 29184811; PMCID: PMC5682371. / Heymann WR. Cutaneous manifestations of thyroid disease. *J Am Acad Dermatol.* 1992 Jun;26(6):885-902. doi: 10.1016/0190-9622(92)70130-8. PMID: 1607406. / Burman KD, McKinley-Grant L. Dermatologic aspects of thyroid disease. *Clin Dermatol.* 2006 Jul-Aug;24(4):247-55. doi: 10.1016/j.clindermatol.2006.04.010. PMID: 16828405. / Kingston ME, Mackey D. Skin clues in the diagnosis of life-threatening infections. *Rev Infect Dis.* 1986 Jan-Feb;8(1):1-11. doi: 10.1093/clinids/8.1.1. PMID: 3513282. / Pulido-Pérez A, Bergón-Sendín M, Suárez-Fernández R, Muñoz-Martín P, Bouza E. Skin and sepsis: contribution of dermatology to a rapid diagnosis. *Infection.* 2021 Aug;49(4):617-629. doi: 10.1007/s15010-021-01608-7. Epub 2021 Apr 15. PMID: 33860474. / Tyler KH. Physiological skin changes during pregnancy. *Clin Obstet Gynecol.* 2015 Mar;58(1):119-24. doi: 10.1097/GRF.0000000000000077. PMID: 25517755. / Mohd Nor B, Wan Razali WMH, Mohd Nor F. Relationship Between Frank's Sign and Cardiovascular Disease: An Autopsy-Based Study. *Cureus.* 2025 Sep 7;17(9):e91756. doi: 10.7759/cureus.91756. PMID: 41063902; PMCID: PMC12501647. / Fernández Ascariz L, Rivas Mundiña B, García Mato E, Limeres Posse J, Alonso Sampedro M, González Quintela A, Gude Sampedro F, Diz Dios P. Frank's Sign and Cardiovascular Risk: An Observational Descriptive Study. *Am J Med.* 2024 Jan;137(1):47-54. doi: 10.1016/j.amjmed.2023.09.019. Epub 2023 Oct 12. PMID: 37832754. / Terziroli Beretta-Piccoli B, Invernizzi P, Gershwin ME, Mainetti C. Skin Manifestations Associated with Autoimmune Liver Diseases: a Systematic Review. *Clin Rev Allergy Immunol.* 2017 Dec;53(3):394-412. doi: 10.1007/s12016-017-8649-9. PMID: 28993983. / Beuers U, Wolters F, Oude Elferink RPJ. Mechanisms of pruritus in cholestasis: understanding and treating the itch. *Nat Rev Gastroenterol Hepatol.* 2023 Jan;20(1):26-36. doi: 10.1038/s41575-022-00687-7. Epub 2022 Oct 28. PMID: 36307649. / Bergasa NV. The itch of liver disease. *Semin Cutan Med Surg.* 2011 Jun;30(2):93-8. doi: 10.1016/j.sder.2011.04.009. PMID: 21767769.

index

Author's acknowledgments

The people I need to acknowledge most for this book on dermatology to be written at all are those who make the bibliography pages, the scientists and researchers and doctors who, over hundreds of years, have dedicated their time to the advancement of knowledge about the skin. Everything we know, every evidence-based decision we make has had a huge team of people behind it, and this book is a compilation of that knowledge. Dermatology in particular is built on meticulous documentation, pattern recognition, history taking and sharing patient stories to spot further patterns. I feel honoured to be able to communicate these to others, but the real hard work in discovering the facts had already been done.

My first direct thanks are to my parents, who really taught me the importance of books and hugely encouraged my love of reading and learning. My parents gave me a love of books so deep that it not only made me want to read as many as possible, but ultimately made me want to write this one. I have loved writing it.

To the doctors who trained me in dermatology, thank you. Thank you for the long nights, for staying a little later with the complex cases, for listening to a patient's story in full, and for teaching me that the story is often the key to the diagnosis and the right treatment for that patient. I owe special thanks to Dr. Du Vivier, who taught me to understand that behind every skin condition is a narrative. The story solves the problem. Thank you for showing me how to find the skin story in everyone and thank you to those patients. And to my greatest pal in medicine, from our first day at medical school together – Dr. Sigi Joseph, the most human and caring GP, always there with words of advice and kindness. I could not have asked for a better friend on this long clinical road.

My gratitude also goes to everyone at DK. The illustrators, the editors, the team – your enthusiasm was contagious. And Phil Hunt, thank you for your encouragement and your gentle reminders of deadlines. You have been a pleasure to work with. Thanks also to my literary agent, the wonderful Amy Thomson – I feel elevated after every meeting with you.

Thank you to Warner Brothers and Full Fat TV, and everyone there who has helped tell the story of dermatology to a global audience. You have brought the patient journey to millions and shown just how profound the impact of skin disease can be on a person's life.

To my lovely and funny husband, Neil – for the past 30 years you have learned more medicine and dermatology than any banker should reasonably be expected to. You have supported me through every endeavour, no matter how many things I say yes to and no matter how many times I have a miniature meltdown when I obviously said yes too many times.

And finally, to my beautiful children, Florence, Saffron, and Rory. Clever, curious, and kind. Thank you for the cups of tea, and helping me to set the tone of this book. You make everything worthwhile.

-)-)-)-

About the author

Emma Craythorne is a leading Consultant Dermatologist and Dermatological Surgeon based at the St. John's Institute of Dermatology at Guy's and St. Thomas' Hospital in London. She specializes in the diagnosis and management of complex skin conditions. Emma serves as an honorary lecturer, teaching both undergraduate and postgraduate students in dermatology, and she delivers specialist training to clinicians across multiple disciplines. Her academic background includes published research in peer-reviewed journals, chapter contributions to major dermatology textbooks, and authorship of her first book, *Pocket Tutor in Dermatology.*

Emma is also known for bringing dermatology to a global audience through her television series *The Bad Skin Clinic* and *Save My Skin.* She has helped demystify skin disease with a combination of scientific precision and deep human empathy. Emma is the founder of www. Klira.skin, a company dedicated to personalized, scientifically grounded skincare, and she is the host of *The After Hours Skin Clinic* podcast.

DK LONDON

Senior Acquisitions Editors Zara Anvari, Kiron Gill
Senior Designer Jordan Lambley
Project Editor Jasmin Lennie
Editorial Assistant Abi Reeves
Senior Production Editor Andy Hilliard
Senior Production Controller Luca Bazzoli
DTP and Design Coordinator Heather Blagden
Jacket and Sales Material Coordinator Serena Sclocco
Art Director Maxine Pedliham
Publishing Director Stephanie Jackson

Editorial Phil Hunt
Design and Illustration Studio Noel
Jacket Design Hannah Naughton

First published in Great Britain in 2026 by
Dorling Kindersley Limited
20 Vauxhall Bridge Road,
London SW1V 2SA

The authorised representative in the EEA is
Dorling Kindersley Verlag GmbH. Arnulfstr. 124,
80636 Munich, Germany

DISCLAIMER

Neither the publisher nor the author is engaged in
rendering professional advice or services to the individual
reader. The ideas, procedures, and suggestions contained
in this book are not intended as a substitute for consulting
with your doctor or a professional. All matters regarding
your health require supervision. Neither the author nor
the publisher shall be liable or responsible for any loss
or damage allegedly arising from any information or
suggestion in this book.

A NOTE ON GENDER IDENTITIES

DK recognizes all gender identities, and acknowledges
that the sex someone was assigned at birth based on their
sexual organs may not align with their own gender identity.
People may self-identify as any gender or no gender
(including, but not limited to, that of a cis or trans woman,
of a cis or trans man, or of a non-binary person). As gender
language, and its use in our society, evolves, the scientific
and medical communities continue to reassess their own
phrasing. Most of the studies referred to in this book use
"women" to describe people whose sex was assigned as
female at birth and "men" to describe people whose sex
was assigned as male at birth.

PUBLISHER'S ACKNOWLEDGMENTS

DK would like to thank John Friend for proofreading,
Vanessa Bird for creating the index, and Aditya Kaytal
for clearing data permissions.

PICTURE CREDITS

101 The Physiological Society: © 2022 The Authors.
Physiological Reports published by Wiley Periodicals LLC
on behalf of The Physiological Society and the American
Physiological Society. / Reference for graph of wound
healing after exercise – Kawanishi, Makoto & Kami, Katsuya
& Nishimura, Yukihide & Minami, Kohei & Senba, Emiko &
Umemoto, Yasunori & Kinoshita, Tokio & Tajima, Fumihiro.
(2022). Exercise induced increase in M2 macrophages
accelerates wound healing in young mice. Physiological
Reports. 10. 10.14814/phy2.15447. https://creativecommons.
org/licenses/by/4.0/ (br)